Dedicated to

My daughter Ziqing, my wife Jing, my PhD supervisor Prof. Fausto Giunchiglia and all my students.

Hao Xu

My grandma, Maria!

Sandro Pinto

My mother, mentors and all my lab-mates in B524.

Yu Xia

My daughter Lara Alicia, my girlfriend POR and all my students around the world, mainly those motivated students at Jilin University and UMinho that somehow challenged us to make them learn in one semester beyond the basic C, and to try new teaching approach that better fits the curriculum purpose, respectively.

Adriano Tavares

"Don't wish it was easier, wish you were better.

Don't wish for less problems, wish for more skills.

Don't wish for less challenge, wish for more wisdom."

Jim Rohn

爱上C语言

徐昊　[葡]桑德罗·平托（SANDRO PINTO）　著
夏瑀　[葡]阿德里亚诺·塔瓦雷斯（ADRIANO TAVARES）

C KISS

Keep It Simple and Straightforward
while playing and enjoying with C
in a programming lab

中国铁道出版社
CHINA RAILWAY PUBLISHING HOUSE

北京市版权局著作权合同登记　图字01-2017-7447

INTRODUCTION TO THE CONTENTS

This book approaches a student-centered C programming teaching and learning, by radically deviating from the classical introductory programming book' organization, writing and content communication as a tutorial. Instead, it covers C language programming how and what as well as the entire life cycle of a C program based on real details of implementation and application, while promoting and mastering debugging skills as the driver of the C programming process. After mastering the C how and what, it ends by smoothly moving from C to C++.

No previous programming background is required as it is intended to be used as the first introductory course in computer programming. A teaching strategy against passive reading is fostered along all classes from the day one, starting first with demonstratives examples followed by hands-on experiences to explore and discover the mechanism of main C concepts. Incentives for active participation in and out of classes will be promoted.

The recommended teaching and learning methodologies for each individual chapter not only leverages a comprehensive and motivational programming training by doing which is supported by the offered code resource, but also enables feedback mechanism to students and instructors toward better improvement of programming knowledge and skills, which will be valuable in future workplaces. The book alongside its suggested methodologies and code resource may be also a valuable training instruction guide for novice enterprise programming professionals.

图书在版编目（CIP）数据

爱上 C 语言 =C KISS:英文/徐昊等著 . —北京:中国铁道出版社 , 2017.9（2018.1 重印）
  ISBN 978-7-113-23978-7

　Ⅰ.①爱… Ⅱ.①徐… Ⅲ.① C 语言 - 程序设计 - 教材 - 英文　Ⅳ.① TP312.8

中国版本图书馆 CIP 数据核字 (2017) 第 269665 号

书　　名：	爱上 C 语言　C KISS	
作　　者：	徐昊　SANDRO PINTO　夏瑀　ADRIANO TAVARES 著	
策　　划：	周　欣	读者热线：(010) 63550836
责任编辑：	周　欣　贾　星　李学敏	
装帧设计：	崔丽芳	
责任校对：	张玉华	
责任印制：	郭向伟	

出版发行：中国铁道出版社（100054，北京市西城区右安门西街 8 号）
网　　址：http://www.tdpress.com/51eds/
印　　刷：北京米开朗优威印刷有限责任公司
版　　次：2017 年 9 月第 1 版　2018 年 1 月第 2 次印刷
开　　本：787 mm×1 092 mm　1/16　印张：13.5　字数：325 千
书　　号：ISBN 978-7-113-23978-7
定　　价：60.00 元

版权所有　侵权必究

凡购买铁道版图书，如有印制质量问题，请与本社教材图书营销部联系调换。电话：(010) 63550836
打击盗版举报电话：(010) 51873659

PREFACE

Introduction

Nowadays, learning a programming language becomes essential to the engineering or science education, as computation is at the core of all modern engineering and science disciplines. *Furthermore, as we also believe in the Digital Technology and AI Era, we postulate that "Nowadays, learning a programming language becomes essential to ALL THE STUDENT, even kids"*. Empowered with programming skills, students will also improve unique and essential problem-solving strategies which are easily transferable to other endeavors. This book approaches a student-centered C programming teaching and learning, by radically deviating from the classical introductory programming book' organization, writing and content communication as a tutorial. Instead, it covers C language programming *how* and *what* as well as the entire life cycle of a C program based on real details of implementation and application, while promoting and mastering debugging skills as the driver or enabler of the C programming process. Leveraging the debugging process at the core of the C programming learning and teaching process will avoid costly future maintenance issues alongside learning autonomy. Mainly, it promotes learners' attitude and spirit towards lifelong learning through learning by doing, as well as by discovering and exploring the *how* of C language by themselves.

Goals and Audience

C was chosen for the introduction to the programming process because it is a procedural programming language and also because we believe engineering and scientific problem-solving is inherently procedural. Additionally, there are several other reasons to choose C as the first programming language such as: (1) to master object-oriented programming concepts students should know first elementary programming concepts, (2) C++, Java and C# share in some way, foundational concepts and syntax of C, (3) C is unbeatable in terms of performance when compared with other high-level programming languages, and (4) major parts of the operating systems like *Windows* and *Linux*, including their device drivers are written in C. This book is written for students with no previous programming experience (i.e., it is intended as a first course in computer programming) and it follows an active and balanced teaching and learning approach for the theory and practice of programming language. It leads students in a holistic manner to ever-improving programming skills based on hands-on experiences and understanding the C programming environment and tools, while avoiding procrastination, enhancing learning autonomy or self-study at the maximum amount of enjoyment.

Our Teaching and Learning Approach

To really empower students with problem-solving and programming skills needed in their future workplace, their learning attitudes and spirit towards motivation, excitement, self-study and resilience, must be presented and enhanced in and after each class alongside hands-on activities. That is, computer programming is not an innate skill, and so, to really become comfortable with the mechanical process of writing and debugging code, students need to practice solving problems and writing code, while simultaneously being willing to learn, not afraid of failing and resilient enough to never give up.

Therefore, a teaching strategy against passive reading should be fostered by teaching not only the concepts of the C language but also the entire life cycle of a C program, while discovering and exploring the how of the main C concepts. All classes from the day one, should be based on hands-on experiences, starting first with demonstratives examples followed by exploration and discovering of the mechanism of main C concepts, as well as pinpointing and fixing program bugs. Incentives for active participation in and out of classes (e.g., doing homework and classroom discussions) should be promoted, mainly for students that are able to understand and discuss programs' execution flow and decide for the best set of breakpoints during a program analysis and exploration. Weaker students should periodically receive feedback regarding their performance while being supported through scaffolding mechanisms, guided by teachers and top performance students.

Key and Unique Features

Contrary to existing introductory programming books that only lightly leverage debugger for bug pinpointing and fixing purposes, this book goes deeper and fosters reverse engineering approach to explore and understand the mechanism of C language concepts. In doing so, a huge degree of excitement and motivation will be triggered, leading students to deeper engagement in their own learning and thus, avoiding procrastination.

Scope of Coverage (Organization and Content)

This book is organized in five chapters, covering C programming language *what* and *how* and ending by smoothly moving from C to C++ and the object-oriented programming paradigm. Chapter 1 briefly presents an overview of C language mechanism, a C programming environment and tools, and the C program life cycle. The main focus of Chapter 2 goes to problem-solving, algorithm and the internal of the function concept in C. Chapter 3 introduces the basics of the pointer variable in C, as well as how does it interoperate with other C language constructs like array and expressions. Chapter 4 focuses on advanced pointers features and how they interoperate with files, functions, passing of arguments, dynamic memory allocation and dynamic data structure. Finally, the focus of Chapter 5 goes towards a brief understanding of the object-oriented programming paradigm, C++ key concepts and refactoring of a C program to C++.

Acknowledgement and Thanks

Besides the authors, many other people influenced this book development and final form. Maybe, the greatest influence came from students who have attended our programming courses for the past years, as they have given great feedback about what works and does not in the classroom. Hence, we specially thank the students for their feedback.

We developed most, but not all, of figures in this book and thus, we would like to thank *Tao Yu* and *Wang Yibo* from Jilin University (China) for drawing some of them.

The reviewers for this book, Prof. Carlos Couto, Prof. Rufino Andrade, Prof. João Monteiro and Prof. Liang Yanchun were very helpful in improving the manuscript, both correcting errors and clarifying the presentation.

CONTENTS

1 Programming in C: an Overview 1

1.1 C Program Development Cycle 2
1.2 Basic Programming Tools and Concepts 4
1.3 Playing with Basic C Language Primitives and Data Types 6
    1.3.1 C Program Skeleton or Structure 7
    1.3.2 Create Your First C Program 9
    1.3.3 Understanding Preprocessor-generated Files 14
    1.3.4 Compilation Process and Error Classes 18
    1.3.5 Playing with Semantic and Syntax Errors 22
    1.3.6 Playing with Runtime Errors: the Debugging Process 25
1.4 Playing with Type Definition, Arrays and Structure 31
1.5 Playing with C Language Constructs 35
1.6 Playing with C Function and Pointer 46

2 Hands-on Functions 54

2.1 Algorithm and Programming 55
2.2 Function Declaration and Definition 62
2.3 Function Call Types and Their Mechanisms 68
2.4 Function and Programming Modularity 72
2.5 Function Scope and Variable Lifetime 81
2.6 Recursive Function 96

3 Hands-on-Pointers: The Basics 104

3.1 Pointer and Memory 105
3.2 Pointer Expressions and Arithmetic 110
3.3 Pointer and Arrays 113
3.4 Pointers and Structures 125

Contents

4 Hands-on-Pointers: Advanced Features 144
4.1 Pointers and Functions 145
4.2 Command Line Arguments: Arguments to Main() 153
4.3 File System Basics and the File Pointer 162
4.4 Dynamic Memory Management in C and Linked List 170

5 From C to C++ 181
5.1 A Brief Overview of C++ Main Features 182
5.2 Usage of Some C++ Key Concepts 184

Index 205

1 PROGRAMMING IN C: AN OVERVIEW

Learning objectives

1. Getting to know basic C language primitives, constructs, functions and pointers and clearly understand what they are without going in too much details.
2. Understanding the meaning of tools and their roles in the programming environment.
3. Mastering debugging capabilities to leverage own learning autonomy out of the classroom.

Theoretical contents

1. C program development cycle.
2. Basic programming tools and concepts:
   a. What is an assembler?
   b. What is a debugger?
   c. Why using a debugger?
   d. What is a compiler?
   e. What is an interpreter?
   f. What is a linker?
   g. What is a preprocessor?
   h. Overview of different kind of file formats.
3. Playing with basic C language primitives and data types:
   a. C program skeleton.
   b. Built-in types.
   c. What are variables and constants?
   d. What is a comment?
   e. Type definition and structure.
4. Playing with C language constructs:
   a. Sequential composition of statements.
   b. Branching.
   c. Loop.
5. Playing with C function and pointer.

Strategies and activities

1. Presenting all C language primitives, constructs and advanced features (e.g., function and pointer) in terms of what they are and what they are not.
2. Demonstrating what primitives, constructs, function and pointer are by explicitly running examples under debugging support.
3. Explicitly showing compiling and linking activities.
4. Running a simple quiz about what primitives, constructs and advance features are and are not.
5. Ending with a demonstration of some acquired debugging skills.
   a. For instance, presenting program with flaws to be pinpointed and fixed individually by students, while explaining the reasons.

Compared to high-level languages, such as Pascal, Ada, BASIC, Python, Ruby or SQL as well as to low-level language as assembly, C/C++ is usually classified as middle-level computer language as it simultaneously leverages the best elements of high-level languages (e.g., richer data types, better portability, better support for software reuse, and natural structure for expressing control flow) and the control and flexibility of assembly language (e.g., manipulation of bits, bytes, words, and pointers). Although providing quite similar control and flexibility of C/C++ regarding the manipulation of bits, bytes and words, Java and C# are also classified as middle-level languages but they do not enable bare-metal manipulations (e.g., pointer manipulation is dynamically managed by garbage collectors on both of them under no programmer control). Such hybridity and its consequently lower abstraction have been promoted C/C++ as the chosen language for the development of system software such as compiler, linker, operating system or hypervisor, where operations on bits, bytes and addresses are very common. C++ language was approached as an extension of C, leveraging more abstract concepts than C.

1.1 C Program Development Cycle

The creation of a program in C or any other high- or middle-level language, usually follows a sequence of steps or phases designated as a program development life cycle. Figure 1.1 presents the program development life cycle followed in this book.

Figure 1.1　C Program Development Cycle: its phases and iterative process towards problem solution enhancement

★ **Problem statement**

A few lines of text which clearly define the problem or objective should be written in this phase to mainly express problem requirements and output of the problem solution. For instance, *"convert odd decimal numbers greater than 20 and lower than 30 to binary and display the sequence of bits on the screen"*.

★ **Analysis**

This phase tries to understand the problem at hand, i.e., what is supposed to be solved. Basically, all solutions' constraints (e.g., 20<n<30 where n is an odd number) and requirements (e.g., input variables of type integer, a function to convert from decimal to binary, and an output variable of type array of integer) should be identified and enumerated in order to solve the problem described above.

★ **Design**

This phase tries to answer how the problem solution will meet the above enumerated requirements, based on the known constraints. Basically, it consists of the identification of subsystems or modules and development of algorithms (e.g., Figure 1.2) to solve the problem as understood at the analysis phase.

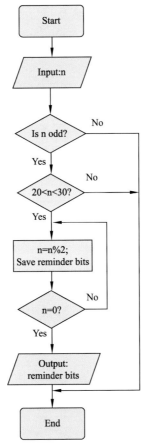

Figure 1.2 An algorithm expressing the decimal2binary conversion for odd numbers greater than 20 and lower than 30

★ **Implementation**

This phase mainly focuses on aspects such as coding and documentation of the problem solution. A programming language, in this case C, is used to write the actual programming statements for each individual steps of the above developed algorithms, producing, for instance, source (i.e., .c) and/or header (i.e., .h) files.

★ **Testing and debugging**

The main focus of this phase is on verifying whether the written code meets the problem requirements under known constraints. Basically, the written code, after being compiled, is tested by inputting several data, including corner cases ones (e.g., negative or even numbers), and check if the desired output is generated. Typically, tests should be performed when concluding each function and modules, as well as when integrating functions and modules in the final program.

★ **Evolution**

Real programs naturally evolves to accommodate new features or functionalities (i.e., changes in requirements) as well as to fix some pinpointed bugs or errors detected after the program deployment and usage. The evolution or maintenance phase should address such required solution enhancements by repeating all previous phases.

1.2 Basic Programming Tools and Concepts

During the *Implementation* and *Testing and Debugging* phases, one usually needs to compile the current written code through a multi-stage process (see Figure 1.3) which can be generically described using the following stages:

★ **Preprocessing**

The first step of the overall build process is carried out by the C preprocessor which is automatically invoked by the compiler to interpret all source file lines starting with the special character # (i.e., each preprocessor command start with #). Basically, what the C preprocessor does is convert the written source code file into another temporary and expanded file with contents referenced by the preprocessor commands. Two of the most used preprocessor commands are the #define and the #include directives, used to define constants and to access externally-defined functions or variables, respectively.

★ **Compilation**

The second step of the build process is converting high-level C statements into lower level instructions (e.g., assembly mnemonics or object code) if there is no error in the source file, and then generating the correspondent "*.obj*" or "*.asm*" file. The compiler can use an integrated assembler to automatically generate object code instead of generating temporarily assembly code that will later be converted to object code by an external assembler. Like a compiler which converts high-level language code, an external or stand-alone assembler converts low-level assembly language code that is microprocessor-specific into an object file.

⋆ **Linking**

The last step of the build process consists in invoking the linker to properly merge object codes in both ".*obj*" and ".*lib*" files (e.g., according to the specified ".*h*" files) to generate the final ".*exe*" file which is specific to a given processor architecture. In so doing, functions initially defined in a specific source file can call other functions or access variables defined in other source files or libraries.

Some compilers (e.g., Visual C++) provide an integrated programming environment (IDE) that already includes an editor and also orchestrates the whole build process by calling internal modules or other programs (e.g., C preprocessor, the compiler itself, assembler, linker and debugger) to handle each stage. Such build process also requires the programmer to work with the following kind of files (see Figure 1.3):

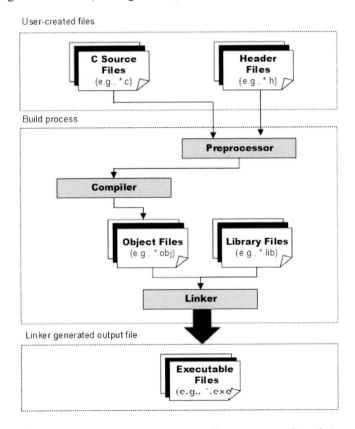

Figure 1.3   *A generic multi-stage build process consisting of Preprocessing, Compilation and Linking stages*

1. Source code files which contain function definitions, and have names which end with ".*c*" extension.
2. Header files which contain function declarations (a.k.a., function prototypes or

signatures[①]) and various preprocessor statements (i.e., lines starting with the special # character). These files end with ".*h*" extension and they are normally used to access externally-defined function (e.g., from the C standard library).

3. Object files are default outputs of the compiler and they consist of function definitions in binary form which may not be executable by themselves as some of them may reference external symbols not known by the compiler. Object files end with ".*o*" or ".*obj*" extensions, being the latter adopted by Visual C IDE.
4. Binary executable files are generated by the linker which link together several object files by mapping previously external unknown symbols to their definitions (i.e., to their currently assigned addresses). Binary executable files end with ".*exe*" extension in the Visual C IDE.
5. Static library files end with ".*lib*" extension in the Visual C IDE and they are automatically handled by the linker to map external unknown symbols (e.g., if ".*h*" files are properly referenced using include directive).

After being generated, a binary executable file can be executed on a microprocessor to generate the problem solution output. Normally, a program execution can be directly executed on the microprocessor from its compiled form, or alternatively be interpreted statement by statement. With few exceptions, C language is naturally compiled but runtime interpreters exist which reads ".*c*" files one line at a time, performing the specific statements contained in that line. In other words, unlike the compiler which processes every statements in a source file at once to be later executed, the interpreter processes and executes on-the-fly the user-written code, line by line. When the execution of a compiled or interpreted program goes wrong due to runtime errors or bugs, a debugger can help you to pinpoint them. A debugger is a program that executes a target program (e.g., the linker-generated binary executable file) and it allows you to monitor, as well as to control the execution of the target program to pinpoint the fault. Basically, it allows you to (1) know which statement was being executed at the instant of the error occurrence, (2) know what are the variables' contents at a specific instant during the execution of the target program, (3) know what is the result of evaluating a given expression at a given instant and (4) know what is the sequence of statements executed in the target program.

1.3 Playing with Basic C Language Primitives and Data Types

Programming languages have been broadly categorized according to two programming paradigms, imperative and declarative, although many more exist (e.g., procedural, structured, functional, object-oriented, event-driven and automata-based), sharing some overlap among them. For example, although most of them were designed to support a specific paradigm, some are flexible enough to leverage multiple paradigms. Mainstream object-oriented languages like C++, C# and Java were designed to primarily support imperative paradigm, while being explicitly extended to also promote functional programming style, e.g., through template metaprogramming (TMP) in C++ or lambda expressions and type inference in C#.

---

① Although signature, prototype and function declaration were defined here as having the same meaning, there are some slight differences among them as compared to the function declaration, signature does not specify the return type (i.e., the function signature consists only of the function name and the parameter list) and prototype can exclude names of formal parameters.

In *imperative or procedural languages,* the programmer focus is on defining computation as statements that change a program's state, i.e., the user-written code follows an algorithmic approach, describing in exact detail the steps that the central processing unit (CPU) must perform to reach a desired state. Basically, the CPU is instructed on how to do something which will result in what the programmer will expect to happen. In *declarative languages* the focus goes towards defining the program logic instead of a detailed execution control flow, i.e., the user-written code will instruct the CPU what the program should accomplish, while letting the CPU to figure out how to achieve it (i.e., without specifying how the program should achieve the result). Functional programming is a form of declarative programming which treats programs as the evaluation of mathematical functions while avoiding state and mutable data.

1.3.1 C Program Skeleton or Structure

Being an imperative language, a C program can be then modeled as a state machine with states defined by values assigned to all variables declared in the program, while the user-written code specifies state transitions (i.e., how program state should change during program execution) by using constructs such assignment, sequence, branch, and loop. Listing 1.1 presents the generic skeleton or structure of a C program which consists of the following parts or sections:

- ★ **Preprocessor directives**

    As briefly said before, these directives are included in the C program source code to instruct the compiler that the source code should be textually changed by calling the C preprocessor before the compile step. All preprocessor directives begin with a pound sign (#) and they all must be on their own line.

Listing 1.1   The generic C program skeleton or structure and its main parts

Preprocessor directives
#include <header_file_1>
...
#include "header_file_n"
#define CNAME value
#define CNAME_1 (expression)
...
#define CNAME_n (expression)
Global declarations
return_type f1(list_of_parameters);
return_type f2(list_of_parameters);
...
return_type fn(list_of_parameters);
datatype variable_1;
...
datatype variable_2;

```
┌─────────────────────────────────────────────────────────────┐
│                    Function definitions                     │
├─────────────────────────────────────────────────────────────┤
│ return-type main(list_of_parameters)                        │
│ {                                                           │
│     Local variables to function main;                       │
│     Sequence of statements associated with function main;   │
│ }                                                           │
│                                                             │
│ return-type f1(list_of_parameters)                          │
│ {                                                           │
│     Local variables to function f1;                         │
│     Sequence of statements associated with function f1;     │
│ }                                                           │
│                                                             │
│ return-type f2(list_of_parameters)                          │
│ {                                                           │
│     Local variables to function f2;                         │
│     Sequence of statements associated with function f2;     │
│ }                                                           │
│                      ⋮                                      │
│ return-type fn(list_of_parameters)                          │
│ {                                                           │
│     Local variables to function fn;                         │
│     Sequence of statements associated with function fn;     │
│ }                                                           │
└─────────────────────────────────────────────────────────────┘
```

★ **Global declarations**

This section contains the declaration of global variables (e.g., *variable_1* and *variable_2*) and user-defined functions (e.g., *f1*, *f2* and *fn*). These global variables have the lifetime of the program, i.e., they can be accessed throughout the program and changes made to them persist to the program lifetime. User-defined functions are declared by their prototype or signature to provide the compiler and also the caller of a function with the basic information regarding the usage of the callee function. The basic information consists of the name of the function, the return type and the list of parameters.

★ **Function definitions**

Definition of a function goes beyond a simple specification of the signature as it also must include the body or behavior of the function. The function body is split into declaration and executable parts, specifying local variables used in the executable part and the sequence of individually semicolon-ended statements to be executed, respectively. Contrary to global variables, local variables have the lifetime of the function call. The bracket sets *()* and *{}* are used to delimit the list of parameters and the body of a function, respectively.

The only function that is mandatory in a C program is *main()* function, which is the first function to be invoked when program execution begins. In a well-written C code, *main()* contains an outline of what the whole program does by calling a set of functions, i.e., acting as a controller function in charge of appropriately (e.g., in their right turn and order) call other functions.

Maybe another useful section is a documentation section, placed above the *Preprocessor directive* section, which is given by a single-line or multiple-line comments with main purpose to provide the name of the program or any other necessary details.

1.3.2 Create Your First C Program

Let's create your first C program by firstly launching *MS Visual Studio* as shown by Figure 1.4.

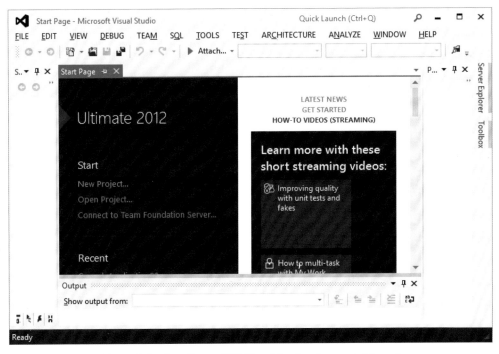

Figure 1.4 Start Page of MS Visual Studio after a quick launch

Secondly, click the *New Project* hyperlink on the left side or alternatively *FILE Menu* at the upper left corner, choose *Win32 Console Application* as shown by Figure 1.5 and then assign a new project name as well as choosing the project directory.

Thirdly, follow the application wizard by clicking the *Next* button (see Figure 1.6) which will give you chance to better customize the project under creation.

For instance, let's create a fully empty project by checking the *Empty project* checkbox while unchecked the "SDL" one as shown in Figure 1.7.

Fourthly, let's create a new source file by right-clicking *Source Files* on the left side and then select *Add* and *New Item…* as shown in Figure 1.8.

Then, choose the default *C++ File (.cpp)* option and rename the source file to your own name with ".*c*" extension as shown in Figure 1.9.

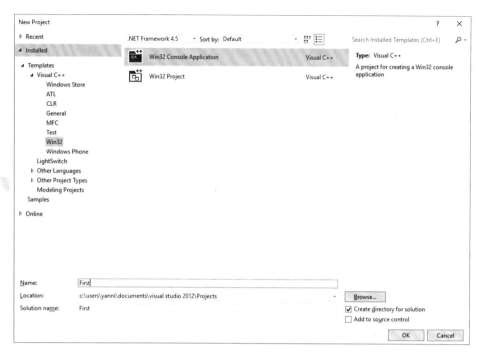

Figure 1.5　Choosing a Win 32 Console Application named "First"

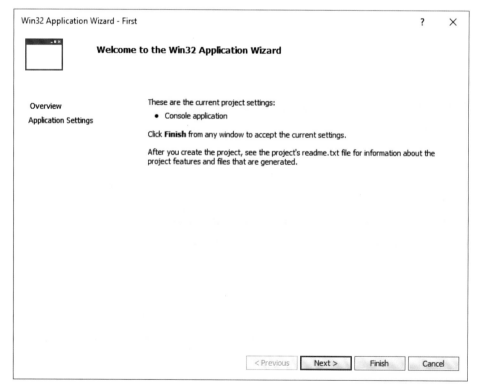

Figure 1.6　Application Wizard form which provides other forms to apply specific settings to the new created project

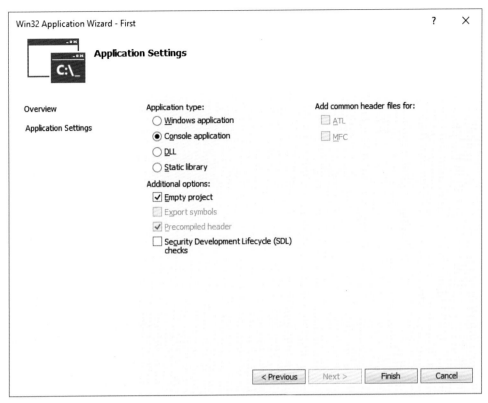

Figure 1.7   Application Setting form which allows custom settings to the created project

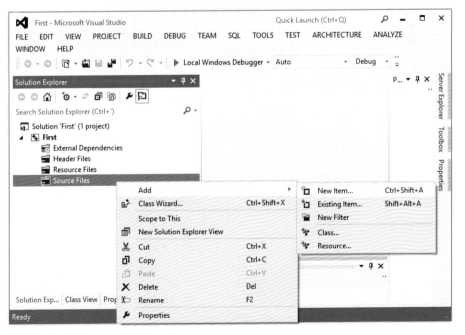

Figure 1.8   First step toward creating an empty source file

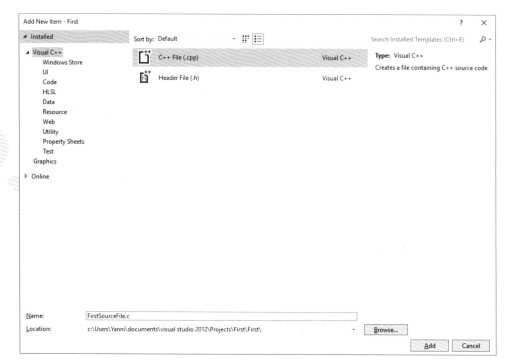

Figure 1.9   Form to choose between the creation of ".h" or ".c", while naming it accordingly

By clicking the *Add* button, the empty source file named "*FirstSourceFile.c*" will be included as part of the new project under the *Source Files* folder (see Figure 1.10).

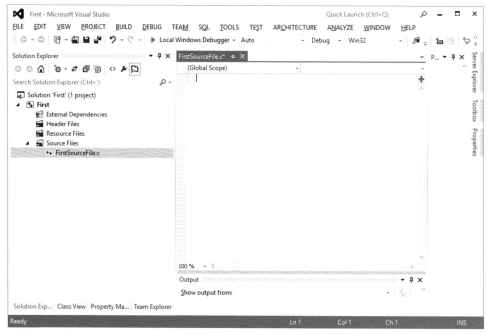

Figure 1.10   Empty source file named FirstSourceFile.c ready to be edited with your own C code

Now it is time to write your first C Program, e.g., like the one shown in Figure 1.11.

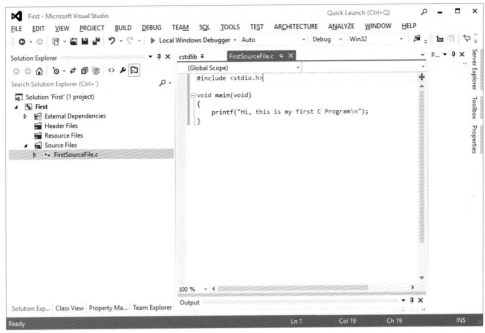

Figure 1.11   A simple C program to display the string "Hi, this is my first C program" on the screen

Fifthly, let's compile it by pressing F7 or Alt+Ctrl+F7 and you will get the following build report shown in Figure 1.12. Alternatively one can also use the window menu under "BUILD|Build Solution" or "BUILD|Rebuild Solution".

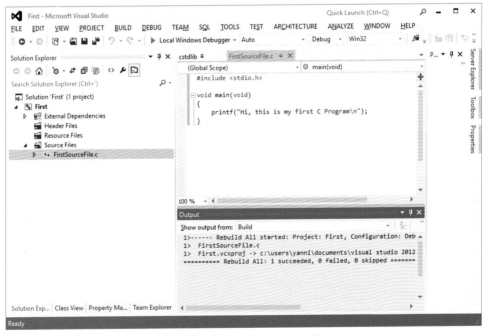

Figure 1.12   Compilation process reporting no error

Finally, you can run or execute the compiled program by pressing Ctrl+F5 and you will get the execution window shown in Figure 1.13. Alternatively one can also use the window menu under "*DEBUG|Start Without Debugging*".

Figure 1.13   Execution windows after running the generated binary executable file named FirstSourceFile.exe

1.3.3 Understanding Preprocessor-generated Files

Now, let's browse through the project folder to see the generated files during the compilation process as shown by Figure 1.14 and Figure 1.15.

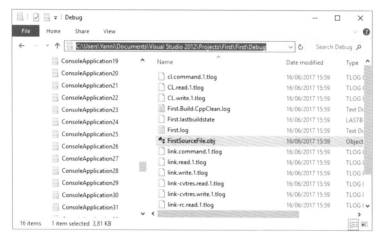

Figure 1.14   The compiler-generated object file

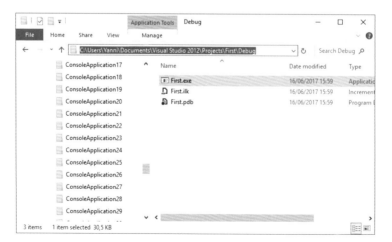

Figure 1.15   The linker-generated binary executable file

Try to figure out what will happen by commenting the preprocessor directive in the above program. Rebuild the modified program by pressing Ctrl+Alt+F7 and then analyze the output window shown in Figure 1.16. In spite of the warning notified an undefined symbol, in this case *"printf"*, the linker generated the *".exe"* file and therefore, you are still able to run it by pressing Ctrl+F5. The point is: how does the linker generate the *".exe"* file, although *"printf"* is an undefined symbol in the program?

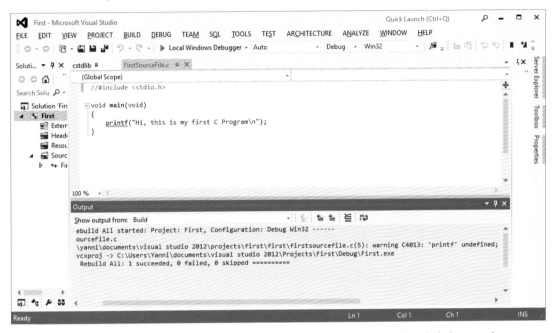

Figure 1.16   Linker detected an undefined external symbol but it was clever enough to link the compiler-generated object file with C standard library

Instead of providing you with keywords for all kind of operations like input/output (I/O), high-level mathematical computation, file handling and character handling, the C compiler usually comes with a standard library of functions, named *libc*, to perform such kind of operations. Calling a library function (e.g., *printf*) forces the compiler to memorize its name to later allow the linker to combine user-written code with object code already found in *libc*. For example, in the above situation, the linker automatically and by default, unless if you instruct the contrary, integrate *libc* object code with compiler-generated object code in the *FirstSourceFile.obj*.

What about functions in *stdlib.h* and *math.h*? Please, try them later to better consolidate your knowledge and understanding of build process internals, as well as the reasons behind the splitting of libm (C standard library for mathematical computations) and libc.

Having an idea about compiler- and linker-generated files, then let's generate and analyze the preprocessor-generated file. Firstly, let's configure *MS Visual Studio* to store preprocessed files (i.e., *".i"*), as by default they are automatically deleted. To do so, the following steps illustrated by Figure 1.17, Figure 1.18 and Figure 1.19 need to be performed.

Step 1: right-click the project name and then select *"Properties"* as showing by Figure 1.17.

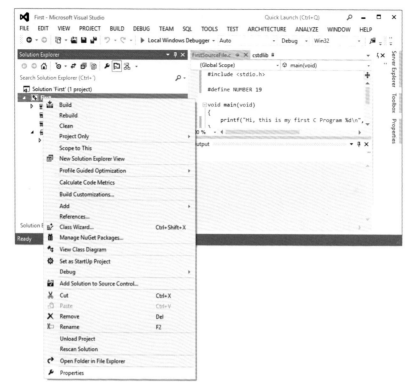

Figure 1.17   Configure MS Visual Studio to store and view the preprocessor-generated files (Step 1)

Step 2: following Figure 1.18, open the "*Configuration Properties*" tab followed by the "*C/C++*" tab and then click on the "*Preprocesso*r" item to change the "*Preprocess to a File*" option to "*Yes (/P)*". Now let's save the configuration by clicking on the "*Apply*" button.

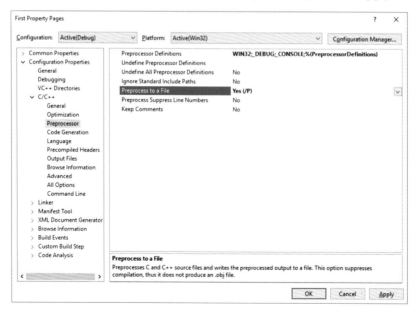

Figure 1.18   Configure MS Visual Studio to store and view the preprocessor-generated files (Step 2 ended by pressing the "Apply" button)

Step 3: call the C preprocessor by pressing Ctrl+Alt+F7 and browse over the project folder to localize the ".*i*" file as showing in Figure 1.19.

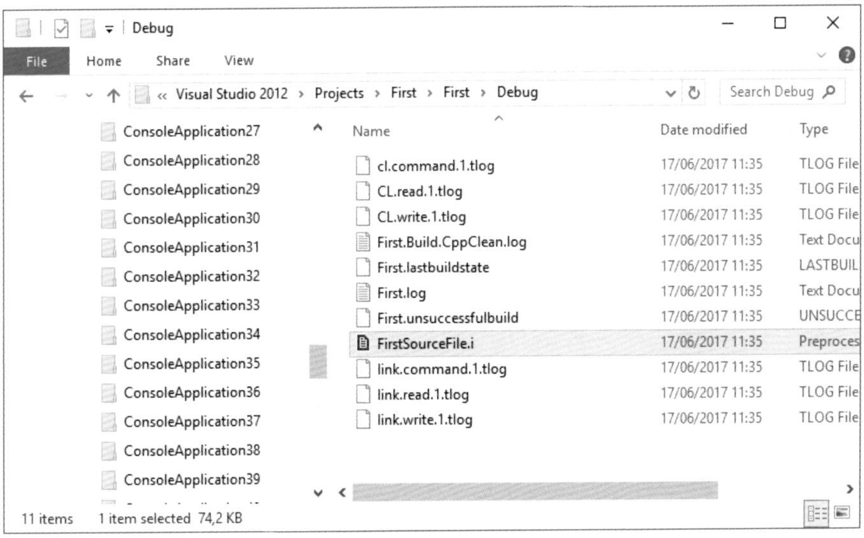

Figure 1.19   FirstSourceFile.i is preprocessed file with expanded information regarding printf() and also the constant NUMBER

Let's briefly analyze the preprocessor-generated file and then restore back the building process configuration to its default settings in order to enable further generation of compiler-generated object files. Figure 1.20 partially shows the "*FirstSourceFile.i*" where the C preprocessor merges both contents of "*stdio.h*" and "*FirstSourceFile.c*" and replaces the constant name by its value while removing both directives from the intermediate preprocessor-generated code. How do you imagine the content of the preprocessor-generated intermediate file if "*#include <stdio.h>*" is commented as shown in Figure 1.21, after pressing Ctrl+Alt+F7?

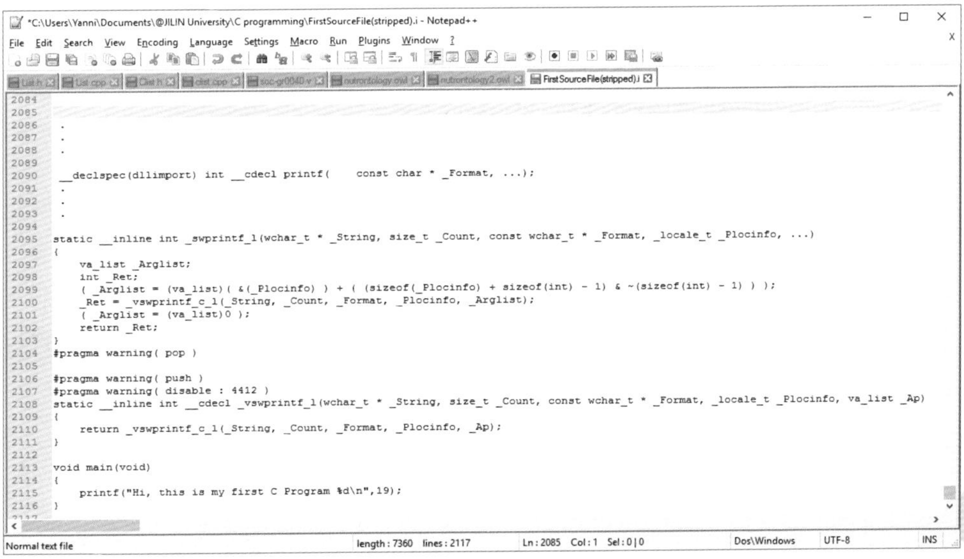

Figure 1.20   A partial view of the preprocessor-generated file

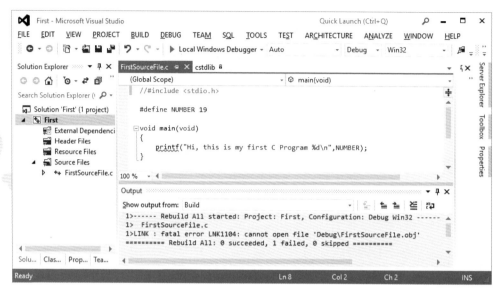

Figure 1.21   Commented #include directive in a build process settings to generated ".i" files. As you can see in the output window, the default settings must be restored to be able to generated object files

As said before, the linker by default integrates *libc* object code with compiler-generated object code, unless you change the default settings of the build process. Therefore, the preprocessor-generated files are completed clean (i.e., without expanded content of *libc* associated to *printf* function) with only the macro expansion to its real value 19 (see Figure 1.22).

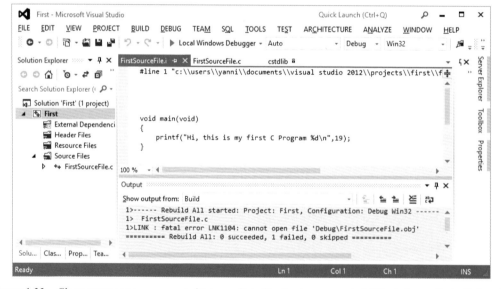

Figure 1.22   Clean preprocessor-generated intermediate file due to commented "#include <stdio.h>" directive

1.3.4   Compilation Process and Error Classes

Let's figure out what will happen by simply renaming the *main(...)*, for instance to *main1(...)*. Reporting to the error on the output window (see Figure 1.23), it will be obvious that *main*() is the only mandatory function in a C program.

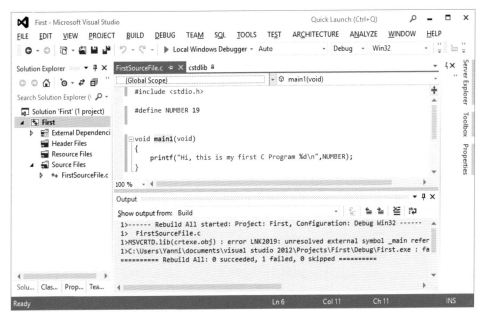

Figure 1.23 Link-time error reported by providing no main(...)

Which kind of error is *"unresolved external symbol _main refer"*? From the three broad categories or classes of error (i.e., compile-, link- and runtime), runtime error can immediately be excluded as you are still in the build process, meaning you are not running the program yet. By the reported error it is obviously a link-time error because the linker is the tool in charge of symbol resolution and relocation, i.e., it determines where external symbols were defined (e.g., other object files) and it maps symbol references to their definitions, respectively. *What could immediately exclude compile-time error?* For instance, browsing over the project folder you will find the compiler-generated "*FileSourceFile.obj*", meaning for a successful compilation step – *remember that the compiler only generated object files if the user-written code has no error.*

Let's then play a little with compile-time errors but before that let's shortly describe how compiler works. Basically, the compilation is done through a series of steps that communicate with one another via temporary files, and are based on the following principles: (1) using several program representations, (2) intermediate representations optimized for various types of manipulation, such as verification and optimization, (3) more dependence on hardware and less on language as it moves along the translation process.

As shown in Figure 1.24, the compilation process is split into several number of relatively independent passes, generically categorized as front- and back-end. Such a decomposition into several and independent passes will be advisable when the source language is large and complex (e.g., a C language), and high quality generated machine code is required. In the front-end, the source code will be processed and an intermediate representation (IR) will be generated after the lexical, syntactic and semantic analysis. The front-end also records in a data structure (i.e., the symbol table) all symbols that appears in the source program by associating to each symbol attributes, such as location, type and scope. A compiler is likely to perform many or all of the following operations:

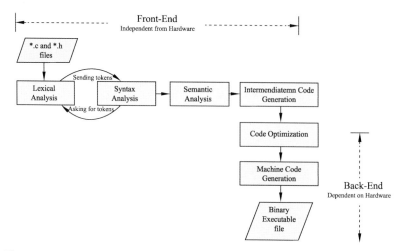

Figure 1.24 The main compile steps: lexical analysis, syntax analysis, semantic analysis, intermediate code generation, code optimization and machine code generation

★ **Lexical analysis (a.k.a., Scanner)**

The scanner reads a stream of characters from ".*h*" and ".*c*" files by splitting them into a set of language tokens (e.g., including C keywords) such as those presented in Listing 1.2. Alternatively, the parser can work in tandem with the scanner, instructing the latter to feed it with tokens while individually analyzing each program statement.

Listing 1.2 Splitting of a C source file according to C language lexical rules

C Program	List of Tokens
int main()	int
{	main
int a=2, b=3;	(
a=a+b;	)
return 0;	{
}	int
	a
	=
	2
	,
	b
	=
	3
	;
	a
	=
	a
	+
	b
	;
	return
	0
	;
	}

★ **Syntax analysis (a.k.a., Parser)**

The parser determines the structure of a program by grouping tokens produced by the scanner into syntactic structures (e.g., usually a parse tree), according to syntactic rules of the languages. Figure 1.25 shows an abstract syntax tree (AST) for the statement "$a = a + b;$", Remember that directives are not statements and so they must be first transformed into statements by the C preprocessor before being passed to the C scanner.

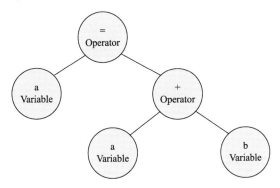

Figure 1.25   Abstract syntax tree for the statement "$a = a + b;$"

★ **Semantic analysis**

In this pass a compiler usually adds semantic information to the parse tree and performs certain checks to enforce program well-formedness based on that information. Basically, this pass tries to capture errors which are not directly detected during syntax analysis. For example, the C programmer is in charge of capturing almost all runtime error due to automatic type conversion or no parameter type checking (e.g., type compatibility between a parameter and an argument) or array boundaries' overrun.

★ **Intermediate code generation**

The main focus of this pass goes toward compiler retargetability, i.e., designing of a compiler that is easily modified to generate code for different CPU instruction set architectures (ISA). Generically, it transforms the parse tree in a more linear IR which is closer to assembly or target machine code, making later the back-end easily changed according to the target CPU.

★ **Code optimization**

Broadly speaking, two optimization classes are usually applied by modern compiler, namely, machine independent and machine dependent optimizations, according to the stage of the compilation process. Nowadays, both kinds try to improve code in terms of memory footprint, power consumption and execution speed. Independently from the stage where the optimization is applied, the program should be semantically preserved, a good tradeoff among optimization metrics should be achieved, and the optimization process should not take too long.

★ **Machine code generator**

This final pass converts a syntactically-correct program into a series of low-level instructions or binary code that could be directly executed by a target CPU. Such transformation is supported on the ISA and runtime environment of the target CPU while it is accomplished through several tasks such as post code generation optimization, selection of instruction, register allocation and ordering of instructions.

Sometimes a pass termed middle-end is used to support compiler retargetability (i.e., usually it performs optimizations that are independent from the source code or machine code) and is intended to enable generic optimizations to be shared between versions of the compiler supporting different languages and target processors. The back-end depends very heavily on the target machine and may perform more analysis, transformations and optimizations than those carried out in the middle-end. Most of those transformations and optimizations are processor and operating system dependent.

1.3.5 Playing with Semantic and Syntax Errors

Looking at the program shown in Figure 1.26, the two local declarations are syntactically valid but they failed a semantic rule dictating that in a given scope (e.g., in this case the main function body) any symbol or identifier should be unique. Notice that the same symbol or name of variable "*i*" is declared twice.

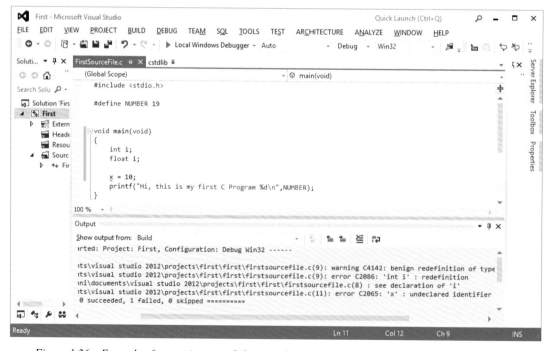

Figure 1.26   Example of semantic error: fail to satisfy symbol unicity and usage only after declaration

The statement "*x = 10;*" also parses fine (i.e., it is syntactically correct) but it fails another of the few C language semantic rules: *any symbol or identifier should be first declared before being used.*

To differentiate between semantic and syntactic errors just remember that syntactic analysis only checks for the program well-formedness while semantic analysis actually checks for the program valid meaning such as bindings, scoping or typing.

So now, let's change the body of *main()* function by commenting previous statements with a multi-line comment form and then typing some new statements (see Figure 1.27). The first reported error is related to a semantic rule restricting the usage of *break* statement only within a *loop* or *switch* control structures. The other two reported errors are caught by type checking variable and recognizing their assigned types (i.e., *int*) do not support the use of field accessor operators like "." and "->".

Figure 1.27   Some more semantics errors

Let's comment all semantically illegal statements and press Ctrl+Alt+F7. Looking at the output window (see Figure 1.28) you will see some semantics warnings that will be fatal or illegal when you will try to execute the linker-generated binary executable file by pressing Ctrl+F5 as shown in Figure 1.29, i.e., they became runtime errors. This program still suffers from more runtime errors that will be later pinpointed and fixed using debugger.

So, what about syntax errors? Let's add two syntactically invalid local declarations at the beginning of *main()* function body and then press Ctrl+Alt+F7. The output window on Figure 1.30 clearly reported that there is no conformant C syntax rules for the added local declarations. First is the misused of operator "+" and the other one is because all C statements should be semicolon-ended.

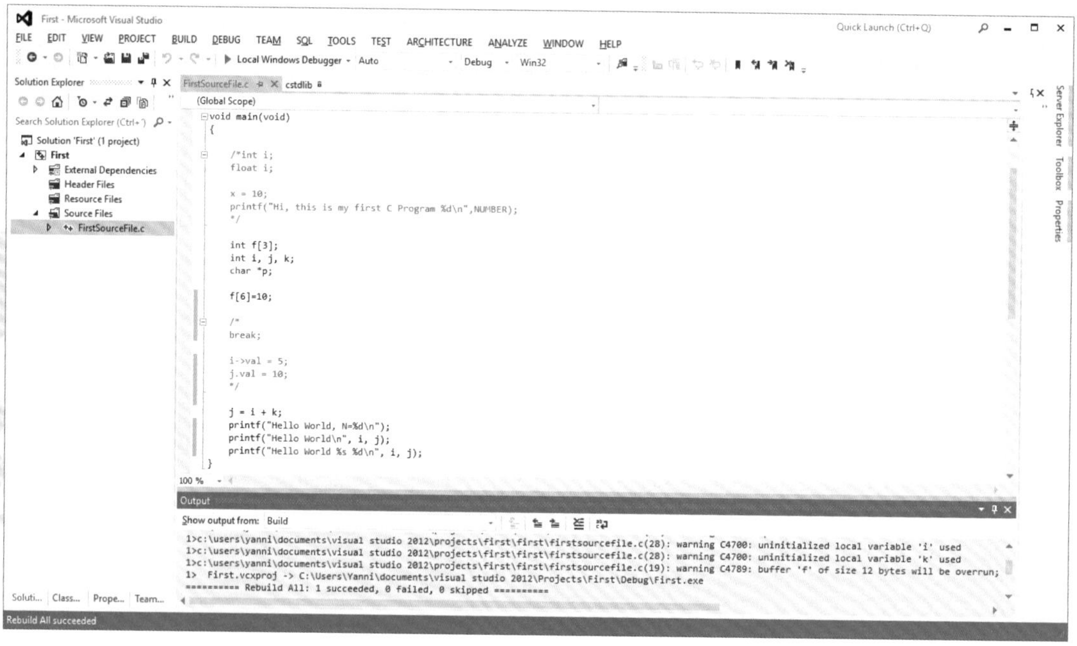

Figure 1.28　Some semantic errors that will become runtime errors

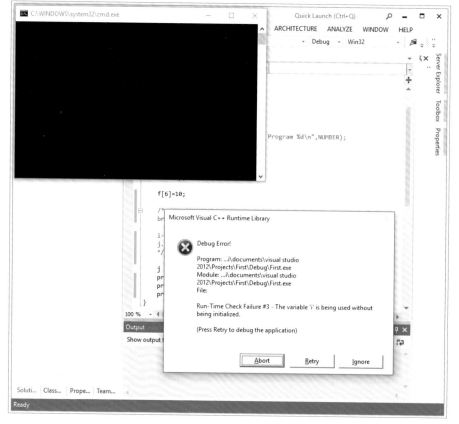

Figure 1.29　Semantic warnings which manifest themselves as runtime errors

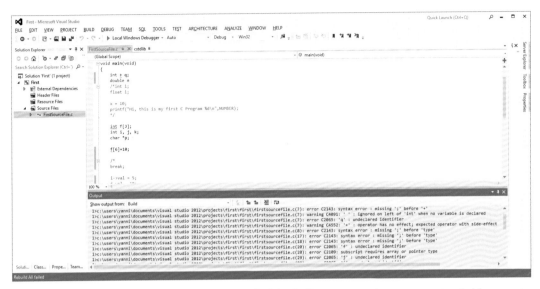

Figure 1.30 Example of two syntax errors: misused of the "+" operator and statement not ended by a semicolon. All other errors is due to the missing semicolon

1.3.6 Playing with Runtime Errors: the Debugging Process

A debugger allows step-by-step execution of the target program, as well as setting breakpoints to temporarily stop the execution at a given program execution point or under specific conditions. When reaching a breakpoint, one can check the state of the program by consulting variables values and then stepping ahead step-by-step to have more detailed information about program execution state. Let's now exercise a little the usage of the debugger by rewinding to the previous program as shown previously in Figure 1.28 and also with execution output on Figure 1.29. Let's then insert two breakpoints, the red circle mark, as shown in Figure 1.31, and then try to fix some of the runtime errors. To insert the breakpoint one can, for example, choose one of the following options: (1) positioning the cursor over a specific statement line and then press F9 or (2) positioning the cursor over the thin gray vertical bar on your left, aligned with the statement line and then press the left mouse button to insert the red circle mark.

Then try to invoke the debugger by pressing F5 and surprisingly, the upper breakpoint was moved down to a different line statement (see Figure 1.32). *Why is that?* The local variable declaration "*int f[3];*" is not an executable statement, and so, it is not eligible for breakpoint setting. However, as a data placeholder it will be eligible for watching.

Let's click on the first two empty entries of the *Watch window* under *Name* column and then write *f* and *i* as the variables to be monitored (see Figure 1.33).

Let's press F10 successively for step-by-step execution while checking the state of each one of the selected variables. After pressing F10 for the first time the execution point is moved to the statement line with "*j = i + k;*", but no change is observed on variable *f*. Rewinding or going back to Figure 1.28 or Figure 1.31, there is a warning reporting an array boundary' overrun which can be dangerous, as maybe the program is trying to write over other variables contents, even belonging to other executing programs. Let's press F5 again and the runtime error will be prompted as shown in Figure 1.34.

Figure 1.31 Setting breakpoints by positioning the cursor on a given statement line and then press F9

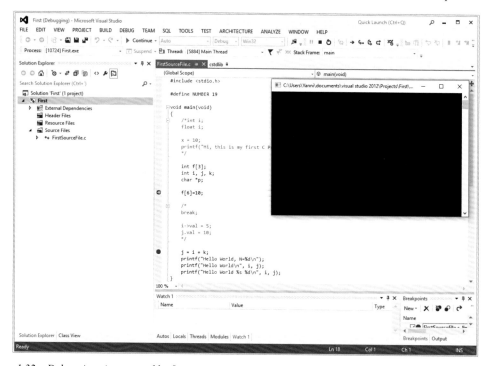

Figure 1.32 Debugging view entered by first setting a breakpoint and then pressing F5 to start the debugging process

1 PROGRAMMING IN C: AN OVERVIEW

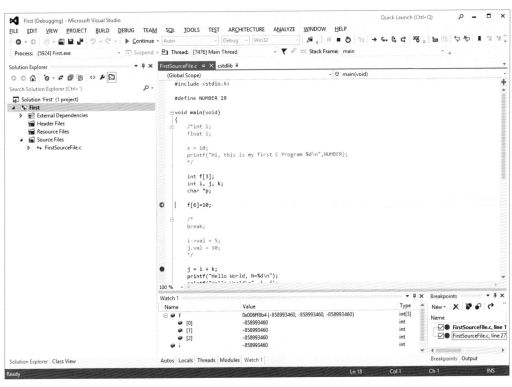

Figure 1.33 Watch Window monitoring the state of variables f and i. While in the debugging mode one can always add new variables

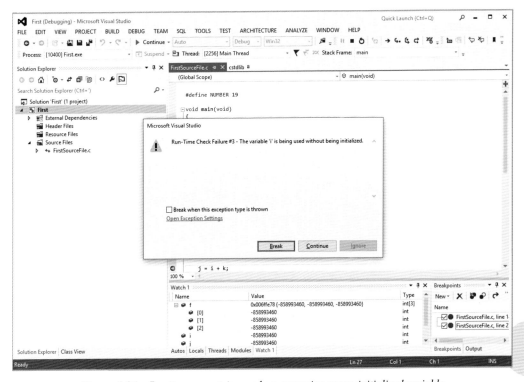

Figure 1.34 Runtime error triggered on accessing a non-initialized variable

Let's press the "*Break*" button and then Shift+F5 to stop debugging and move to the editing mode. Let's insert two new assignment statements to initialize the variables *i* and *k*, repeat the whole build process, enter the debugging mode, update *Watch window* for monitoring the state of *i* and *k*, insert a new breakpoint at the last statement, and finally proceed with continuous execution mode by pressing F5 twice (see Figure 1.35). As you can see the runtime error on the statement at the second breakpoint was already fixed while you can also observe the current state of the program execution, as given by the *Watch window* and the execution status on the DOS-like command line window.

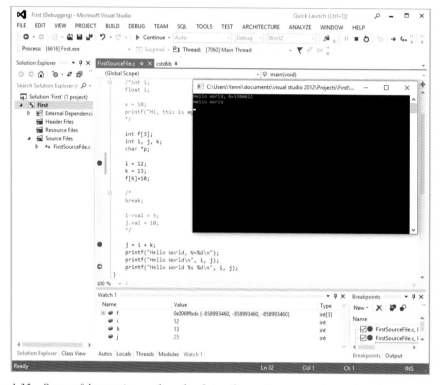

Figure 1.35 Successful execution, so far, after fixing the runtime error triggered on the statement at the second breakpoint

Can you guess what will happen next by pressing F10 or F5? Another runtime error will be triggered as is shown by Figure 1.36, mainly because C semantic analyzer does not check for type compatibility between a parameter and its corresponding argument, while it also promotes automatic type conversion of arguments to their corresponding parameter types. While executing the call of *printf ()* at the last breakpoint, the argument *i* is automatically converted to a string of character as specified by the *%s* format specifier, resulting in memory access beyond the size of the *int* type of the argument *i*. Looking at the *editor window* you will see that the program execution jumped to code on *libc* where the type compatibility check was supposed to be done.

Once the error is pinpointed, you can fix it by changing the format specifier from *%s* to *%d* and repeat again all the previous steps from build to debug processes (see Figure 1.37).

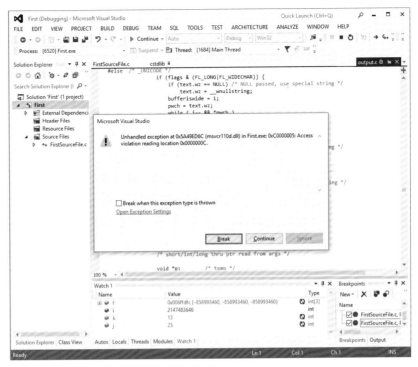

Figure 1.36   A new runtime error is triggered because C left to the programmer the duty for checking type compatibility between a parameter and an argument during the call of a function (e.g., printf in this case)

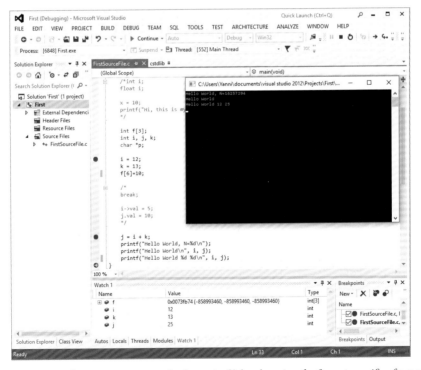

Figure 1.37   Fixing the runtime error on the last printf() by changing the format specifier from string of character to integer

However, forcing another debugging step through step-by-step or continuous execution you will realize that the program is still suffering from another access violation error. *Why is that?* The error is due to the array boundary' overrun on the "*f[6]=10;*" . Therefore, change the index from *6* to *2*, repeat the process till executing such statement and you will observe now the successful change of the state of variable *f* (see Figure 1.38).

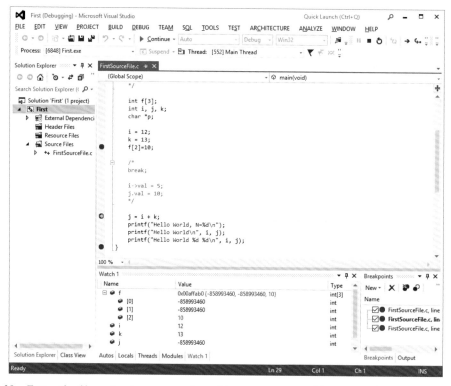

Figure 1.38   Fixing the f boundary' overrun and watching now a successful access to the array on index 2 with value 10

Repeating continuous execution till the end, you will reach a successful debugging of that program. Now, you can execute again the program without debugging by pressing Ctrl+F5 and you will see the following execution window on Figure 1.39.

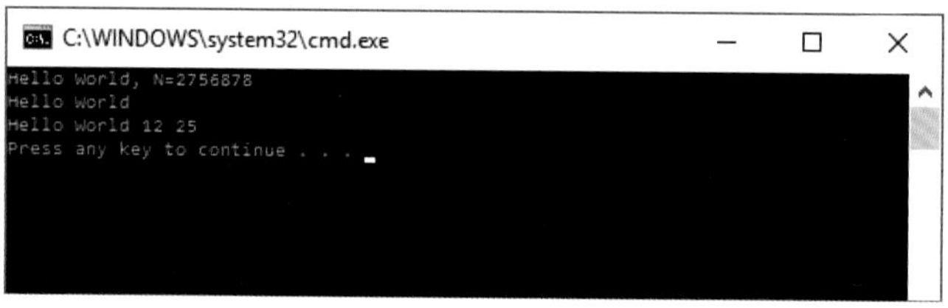

Figure 1.39   Execution window after successful execution

Other minor semantic errors which do not manifest as runtime errors are still presented on first two *printf()* calls. *Can you explain them?*

1.4 Playing with Type Definition, Arrays and Structure

Based on the above experiments, variables and data types could shortly be defined as named locations in computer memory (i.e., memory addresses) that usually hold values or state of programs and representations of data on computer memory, along with the set of operations that can be performed on corresponding declared variables, respectively.

As you previously saw, more than simply satisfying syntax rules for variable declaration in the form of "*type_name list_of_variable_names*;", semantic rules like "*declaration before usage*" and "*usage without being previously initialized*" must also be met.

C language provides several built-in types such as character (i.e., *char*), integer (i.e., *int*), floating-point (i.e., *float*), double floating-point (i.e., *double*), and valueless (i.e., *void*), whose size and range in memory varies among CPUs and compilers, except for a variable of type *char* that is always one byte. The size of a variable of type *int*, is in the compiler parlance, dictated by the *machine word* which is based on the bus width of the target CPU, while the exact format of variables of type *float* and *double* will depend upon the method used to represent them. Let's use now the *sizeof* operator to exercise a little on the sizes of built-in data types as shown in Figure 1.40. This operator will provide a way to avoid specifying machine-dependent data sizes in a program.

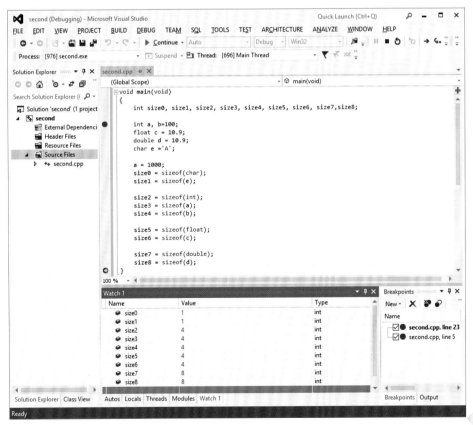

Figure 1.40 Exercising with sizeof operator to determine the amount of storage in bytes needed to represent variables or datatypes

Now that you know about built-in types, what about creating your own custom data type (i.e., user-defined type) in C? The following approaches are available for the effect:

1. *struct* statement which defines new data types by grouping together several already existing members or fields types (i.e., variables) under one single name (see Figure 1.41).

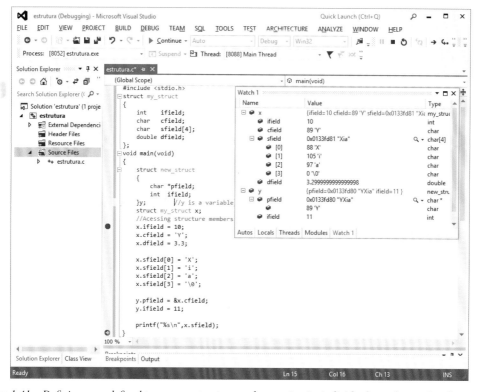

Figure 1.41   Defining user-defined types as a structure and accessing its individual members using dot operator

2. *typedef* keyword which creates alias names for user-defined and built-in types, i.e., it assigns a new name to another already existing type. It is mainly used for two purposes: (a) to simplify declaration of pointer type or aggregate types based on *struct* and *union* (see Figure 1.42), as well as (b) to declare context specific types by explicitly indicating the meaning of a variable definition (see Figure 1.43). *Typedef* keyword is ruled by the following syntax format: "*typedef existing_type alias_type_name;*" or simply "*typedef typedeclaration;*".

3. *union* statement which enables the same memory space to be defined as different types of variables.

Through step-by-step debugging, you will see that *cfield* and *sfield* members will be first correctly modified, but later both got corrupted by the final value of *cfield1* member, which will be at the end the only one preserving the exactly written value (see Figure 1.43). As a conclusion, *union* members should be used only one at a time as simultaneous usage corrupted previous used members. To conclude, Listing 1.3 shows the syntax format for the *union* statement where both *union_tag* and *union_variable* are optional. The *struct* statement also shares the same syntax by only exchanging the *union* word with *struct*, mainly the initial *struct* statement keyword.

1 PROGRAMMING IN C: AN OVERVIEW • 33

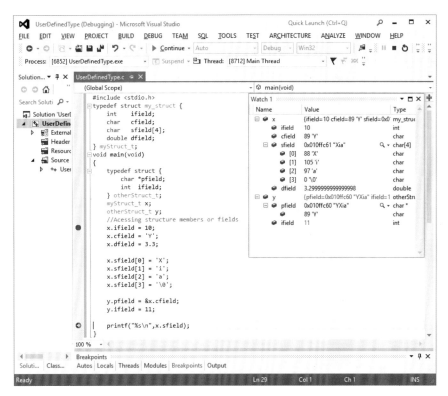

Figure 1.42   Using typedef to simplify declaration of variables x and y through the elimination of the keyword struct

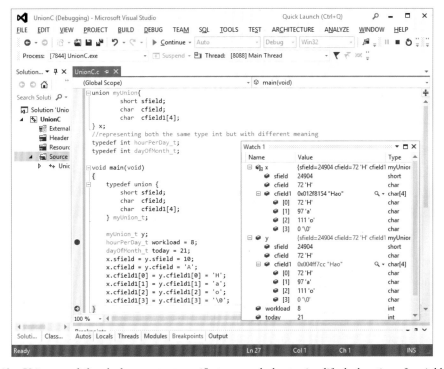

Figure 1.43   Using typedef to declare context specific types and also to simplify declaration of variables x and y through the elimination of the keyword union

4. *bit-field* statement specifying a special type of structure or *union* element which allows bit manipulation. Figure 1.44 shows how to save some memory space as you do not need for more than 1-bit and 3-bits to represent binary and octal numbers, respectively.

Guess what will happen if you write values requiring more bits representation than the specified width of a bit-field members? Please try it by yourself. Try also to replace the tag on the global structure by an empty tag and see what will happen. Listing 1.4 shows the syntax format for the *bit-field* statement.

Listing 1.3   Union statement format

```
union [union_tag]
{
    member definition;
    …
    member definition;
} [union_variables];
```

Listing 1.4   Bit-field statement format

```
struct [bit-field_tag]
{
    type member_name : width;
    …
    type member_name : width;
} [bit-field_variables];
```

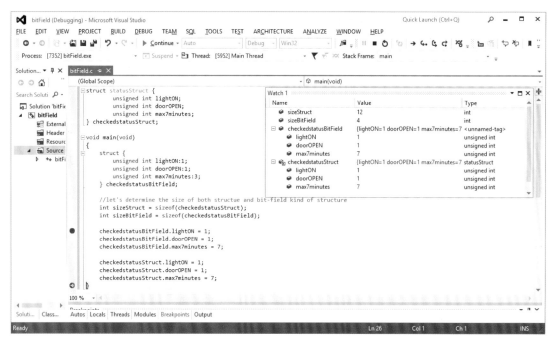

Figure 1.44   Using bit-field to save memory space but the point is: at what cost?

5. *enum* statement specifying list of named integer constants all known ahead of time.

From Figure 1.45, it is visible that by default the value of the first constant is zero, unless explicitly assigned. It is also visible that the next constant will be exactly an increment by one of the previous one, if it was not explicitly assigned. Listing 1.5 shows the syntax format for the *enum* statement.

Finally, let's shortly play with an array where each element is declared through a structure-based and user-defined type as shown in the following Figure 1.46.

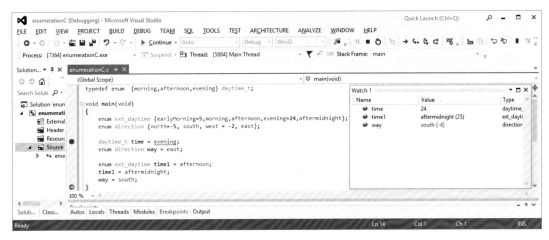

Figure 1.45   Exploring enumeration type and its initialization.

Listing 1.5   syntax format for the enum statement

enum [enum_tag] { comma-separated named constant list} [enum_variables];

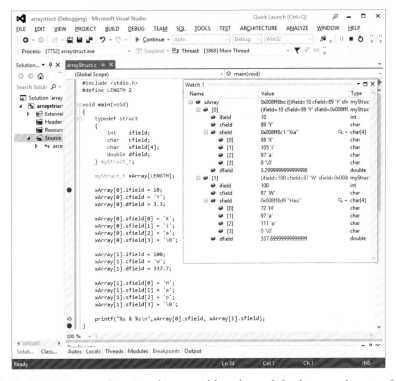

Figure 1.46   Exploring a combination of structured-based user-defined type as elements of an array

1.5   Playing with C Language Constructs

Imperative languages like C, use constructs (a.k.a., control constructs) to control the flow in which statements are executed or not. So far, all previously written and executed programs run sequentially as none of them behaviorally presented control constructs statements. Basically, a program execution flow is described by combining the following four control primitives, namely selection, sequence, repetition and/or function invocation, which are also statements on their own.

By default, any regular statement (i.e., assignments, declarations but never a control construct) in imperatives languages is a sequence construct, instructing the CPU about the next statement to be executed. Usually, it is the first statement in the program or the statement following the current one. A special statement given by the keyword *goto* can be used to change the regular sequential execution, consisting of a set of only semicolon-separated regular statements.

C as an imperative language provides two types of selection statement represented by *if* and *switch* statements. Figure 1.47 shows *if* statement syntactically and algorithmically, while Figure 1.48 demonstrates how to play with a program execution flow according to the state of the Boolean expression associated to *if* statement. To trigger the block corresponding to the false condition (i.e., the else block), please change the variable *cond* to any number different than *1* and then invoke again the debugger in a step-by-step mode. Remember also that the *else* block can be optional.

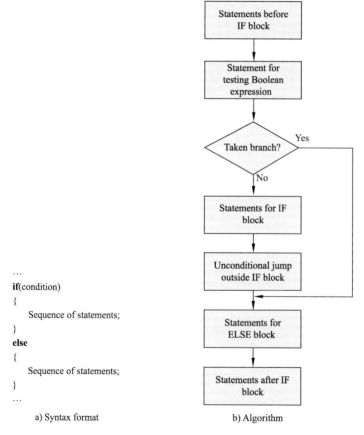

Figure 1.47 IF selection construct

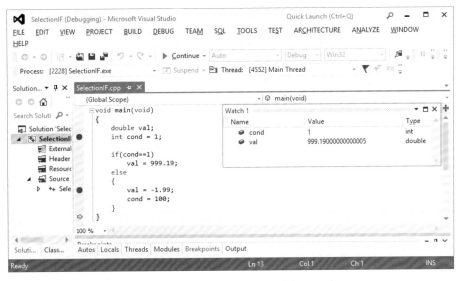

Figure 1.48 Firing the debugging of if block of the IF selection construct

Contrary to *if* statement which supports selection only from two alternatives (i.e., unless if they are nested), *switch* case statement allows selection from many alternatives where each one is linked to a selector (see Figure 1.49). When the selector evaluates to true it triggers the execution of an associated sequence of statements. Figure 1.50 and Figure 1.51 show the step-by-step debugging execution mode by inputting values 7 and 3 to the *selector* variable. *Guess what will be the value of x by inputting the value 2*. Please notice that the *break* statement prevents the code to continue executing into the next case.

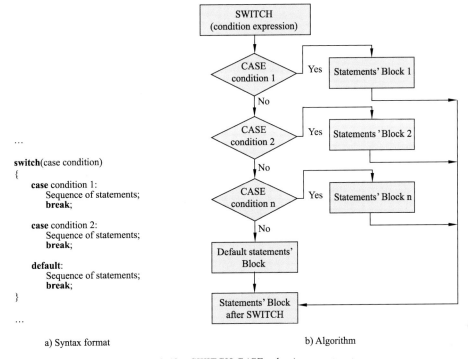

Figure 1.49 SWITCH CASE selection construct

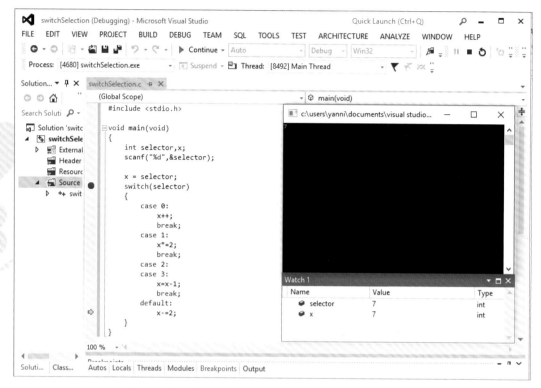

Figure 1.50　Firing the default block by entering selector as 7

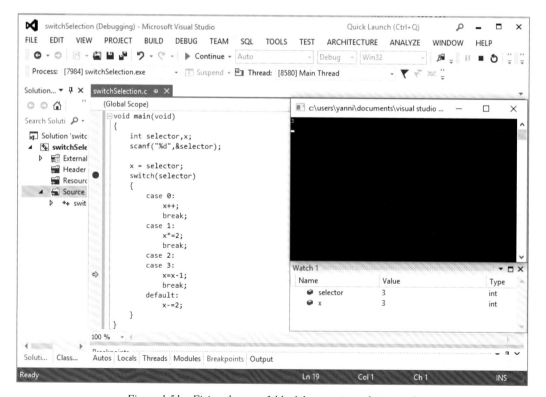

Figure 1.51　Firing the case 3 block by entering selector as 3

Can you figure out to which data types are the selector variable constrained to? Since *int* was already tried, please try other types (e.g., starting by *char*, *enum*, *float* e.t.c.) and repeat the build process to identify corresponding illegal types. *Should case statements be presented in a correct numeric order?* Please change the constants in the first two cases and retry both the build and debugging processes.

Using the previous example in Figure 1.51, let's: (1) insert a breakpoint at the statement "*scanf("%d",&selector);*", (2) start the debugging process in step-by-step into mode (i.e., pressing F11) instead of a step over as done so far by pressing F10, (3) execute *libc* code by pressing F5 to return from *scanf(...)* directly to the next breakpoint, and voilà. Figure 1.52 illustrates step 2 by entering the *libc* code. By pressing now F5 you'll be asked to enter the selector value and then you will get back to *main()* function code, exactly at the second inserted breakpoint (see Figure 1.53).

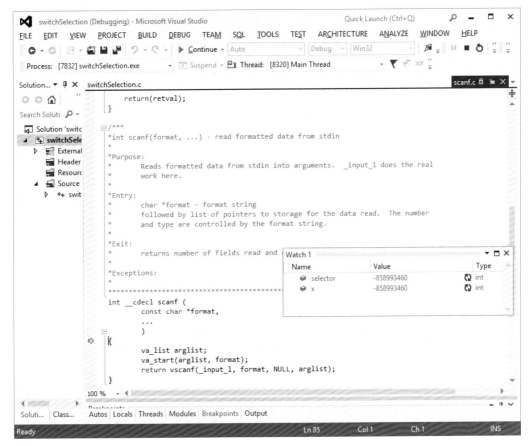

Figure 1.52   Function call construct: getting into the libc function with step into debugging mode (F11)

Figure 1.54 illustrates the function call process which seems very similar to the sequence construct given by the *goto* statement, except that it guarantees that the execution flow will return back to that point from where the routine was called.

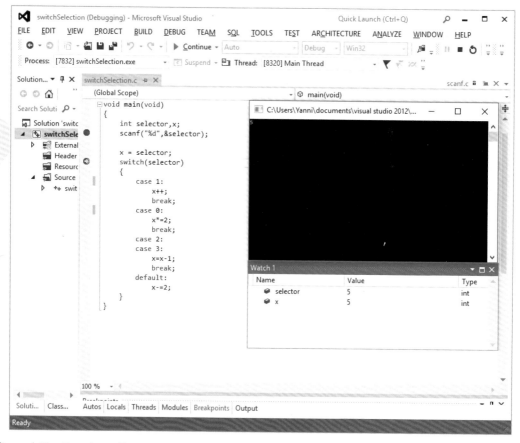

Figure 1.53 Function call construct: returning to the main() function at the 2nd breakpoint after pressing F5, inputting the selector value and pressing ENTER

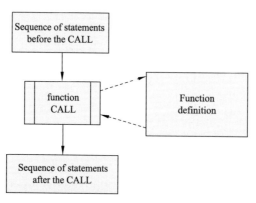

Figure 1.54 Graphical flow of function call construct. Notice the new graphical symbol representing the function call instead of the regular graphical symbol used so far for representing sequence of statements

Let's continue the programming journey toward *repetition* construct, which repeatedly causes execution of a sequence of statements until some end condition is met. In C, it can be one of the following three types: *while*, *do while* and *for* statements.

Let's start with *for* statement as represented syntactically and algorithmically by Figure 1.55.

Figure 1.56 illustrates debugging sequence regarding to full repetitions till the exit from the sequence of statements, while Figure 1.57 shows the start of execution for the second iteration to show how the *update* statement will affect the state of variable *i*.

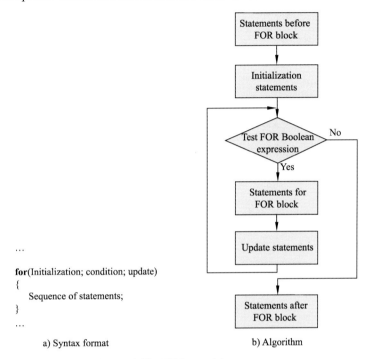

Figure 1.55 FOR repetition construct

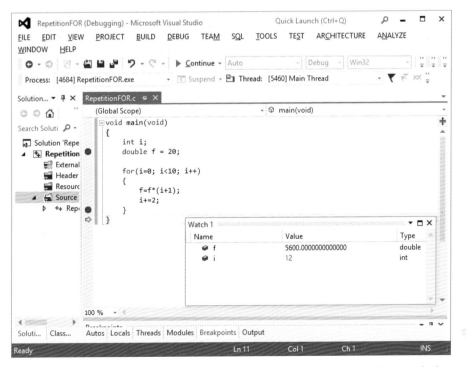

Figure 1.56 FOR construct: debugging sequence regarding full repetitions till exiting the loop

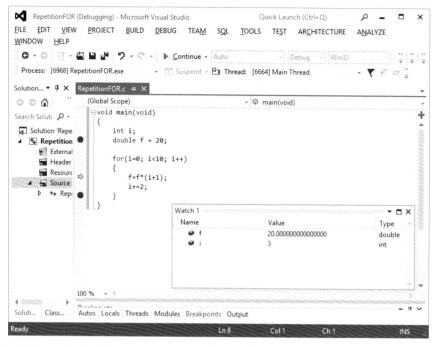

Figure 1.57   FOR construct: debugging sequence at the very beginning of the 2nd loop iteration to understand the effect of the sequence of update statements

Figure 1.58 represents the *while* construct syntactically as well as algorithmically, while Figure 1.59 illustrates the debugging sequence regarding to full repetitions till the exit from the execution loop.

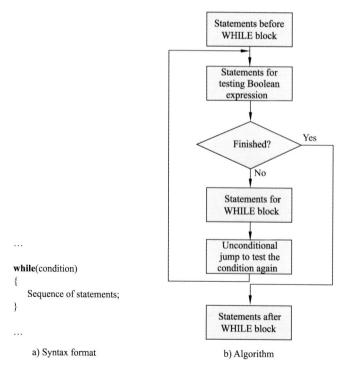

Figure 1.58   WHILE construct

1 PROGRAMMING IN C: AN OVERVIEW • 43

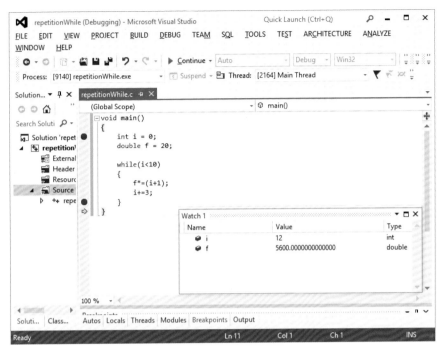

Figure 1.59 WHILE construct: Debugging sequence regarding full repetitions till exiting the loop

Finally, let's exercise with the *do ··· while* construct which can be syntactically and algorithmically represented as shown in Figure 1.60.

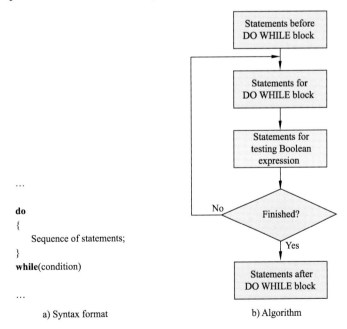

Figure 1.60   DO ··· WHILE construct

Figure 1.61 illustrates the debugging sequence regarding full repetitions of the loop sequence of statement until the exit from the loop, due to the condition failure.

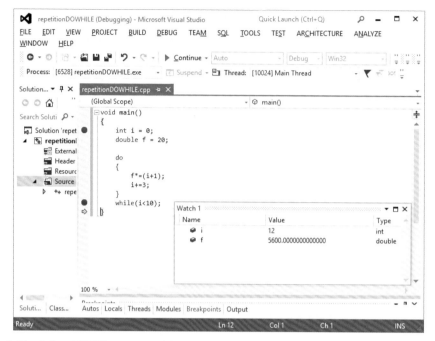

Figure 1.61 DO ⋯ WHILE construct: debugging sequence regarding full repetitions till exiting the loop

Maybe sometimes and exceptionally one would like to exit the loop under a more specific condition. That is exactly when the *break* statement, as algorithmically described in Figure 1.62, comes up.

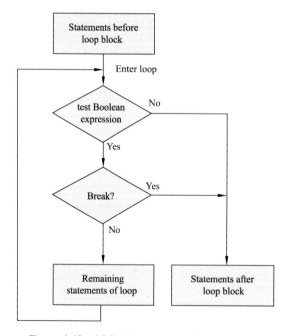

Figure 1.62 BREAK statement and its algorithm

What if you want to exit the loop as soon the value of *f* becomes greater than 500? Figure 1.63 illustrates the debugging sequence regarding to the execution till the failure of the internal more specific condition fail (i.e., *f<=500*).

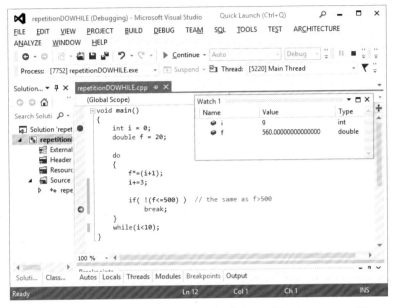

Figure 1.63 BREAK statement: debugging sequence till the failure of the internal condition and consequently exiting the loop

However, most of the time there is a better alternative to the usage of the *break* statement as illustrated by the solution in the following Figure 1.64. The internal condition attached to the *break* statement was first inverted using the logical negation operator and then combined with the previous external one, forming a new compound external loop exit condition.

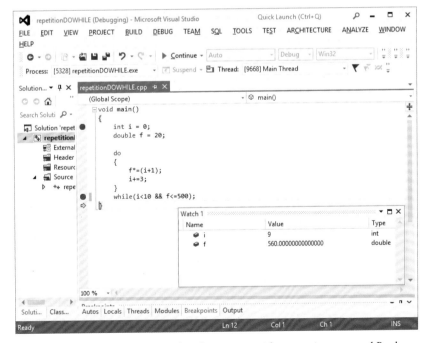

Figure 1.64 Alternative solution replacing break statement with a generic compound Boolean expression

1.6 Playing with C Function and Pointer

A function is a sequence of statements forming a conceptual unit about a specific task to be performed, and so, it will become useful to understand its separation and individuality. In general, it encapsulates the sequence of statements, including the *return* statement, in a named function, as syntactically shown in Listing 1.1, which will be then replaced with a function call in the *main*() or any other function (see Figure 1.54 for function call construct).

As discussed above, programming in any language, goes beyond coding-only as done, for example, at the implementation phase, without being preceded by analysis and design phases of the C Program Development Life Cycle (see Figure 1.1). Therefore, let's follow the whole proposed approach and continue with the implementation of the algorithm for the decimal to binary problem as presented in Figure 1.2. Figure 1.65 presents the user-written code while illustrating debugging sequence and watching state inside the body of the function (i.e., debugging the function behavior). Figure 1.66 presents the debugging sequence for the *main()* while calling *dec2binOdd()*.

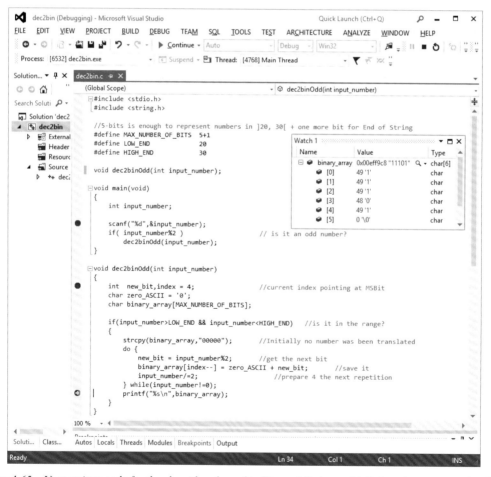

Figure 1.65 User-written code for the algorithm showed in Figure 1.2 along with Debugging sequence inside the body of dec2binOdd()

1 PROGRAMMING IN C: AN OVERVIEW 47

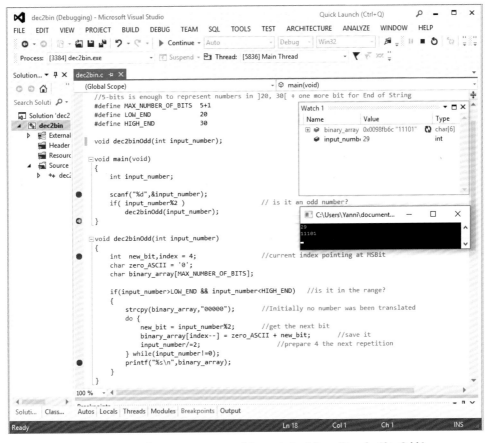

Figure 1.66 Debugging sequence of the main() while calling dec2binOdd()

For better and more detailed understanding of a C program, Table 1.1 presents a dissection of the program played with in Figure 1.65 and Figure 1.66.

Table 1.1 Dissecting line by line the Dec2bin program

C Program	Comments
#include <stdio.h>	stdio.h is the standard header file with the function prototype for *printf( )* and *scanf()*
#include <string.h>	string.h is the standard header file with the function prototype for *strcpy( )*
#define MAX_NUMBER_OF_BITS 5+1	This symbolic constant sets the size of the character array for the 5 converted bits. According to the range of the number to be converted 5 bits is enough to representing them all. One more bit is needed for ending the string with '\0'
#define LOW_END 20	This symbolic constant sets the lower-end of the range interval of convertible numbers
#define HIGH_END 30	This symbolic constant sets the higher-end of the range interval of convertible numbers
void dec2binOdd(int input_number);	Function declaration for *dec2binOdd()*, informing the compiler about the name as *dec2binOdd*, which is a function that returns no value and receive one input parameter of type integer named *input_number*
void main(void)	Starts the definition of the *main()* by including its declaration as a function that returns nothing and also receives no argument
{	Begins the body of *main()* with the left bracket symbol

Continued

C Program	Comments
int input_number;	Declares an integer local variable named *input_number*
scanf("%d",&input_number);	*scanf()* is waiting for an integer number to be entered by the user and stores it in the location assigned to the variable named *input_number*
if(input_number%2)	Check if the read number on *input_number* is odd by analyzing the remainder of division
dec2binOdd(input_number);	Converts if it is an odd number by invoking *dec2binOdd()*
}	Ends the body of *main()* with the right bracket symbol
void dec2binOdd(int input_number)	Starts the definition of *dec2binOdd()* by its declaration as a function returning no value and taking one integer input
{	Begins the body of *dec2binOdd()* with the left bracket symbol
int new_bit,index = 4;	Declares two integer local variables *new_bit* and *index* and initializes the latter with 4
char zero_ASCII = '0'; char binary_array[MAX_NUMBER_OF_BITS];	Declares a character local variable *zero_ASCII* and initializes it with the representation of zero in ASCII, while creating an array named *binary_array* as a container for 6 characters for the converted bit plus the end of the array
if(input_number>LOW_END && input_number<HIGH_END)	Checks if the number is in the required range
{	Opens the multiple statements *if*-block with left bracket symbol
strcpy(binary_array,"00000");	If the read number satisfies the problem constraints, initially sets all bits in *binary_array* as '0'
do {	Opens the multiple statements *do···while* block with the left bracket symbol
new_bit = input_number%2; binary_array[index--] = zero_ASCII + new_bit;	Gets the new converted bit as the remainder of division by 2, stores the value on *new_bit* and finally converts the integer representation of the bit to ASCII to be stored at the next most significant available bit position as indicated by the current variable index. Updates index for the next bit position
input_number/=2;	Updates the conversion process by dividing the content of *input_number* by 2
} while(input_number!=0);	Closes the multiple statements *do···while* block with the right bracket symbol while checking for the ending condition. If *input_number* is different from zero goes back and repeats the *do···while* block execution
printf("%s\n",binary_array);	Prints on the screen the converted bits
}	Closes the multiple statements *if*-block with the right bracket symbol
}	Ends the body of *dec2binOdd()* with the right bracket symbol

Both debugging sequences were driven by the continuous execution mode (i.e., by repeatedly pressing *F5*), but they could also be driven through step-by-step debugging mode by repeatedly pressing *F10* and *F11*, with the latter used to force debug entry into function body. *Please try it by yourself.*

Looking at the code you will see how the *do ··· while* construct graphically presented in the algorithm at Figure 1.2 was translated to C *do ··· while* construct. It also shows the usage of the function call construct in the *main()* function where the function parameter (i.e., *input_number* of type integer) as presented in the function declaration was replaced at the calling point by the argument (a.k.a., actual parameter which is the content of variable *input_number*, in this example, the value 29). However, some improvements can be made to the written code, for instance, replacing the local variable declaration given by the statement "*char zero_ASCII = '0';*" with a preprocessor directive for constant definition. *Please try it by yourself and also exercise your coding skill by providing the same function implementation using oth-*

er control constructs and data types. Before, try to deeply debug the code by inputting corner cases values like those out of the range (e.g., enter numbers greater than 29 or lesser than 21) and draw the followed debugging sequence.

Before trying to answer *what a pointer is*, as well as briefly exercising with it, let's first summarize with the debugger help (see Figure 1.67) what a variable is by identifying its five main attributes:

(1) A variable has a type (e.g., *int, char, double, struct testStruct* or array of character);
(2) A variable has a name (e.g., *size, i, c, d, str* or *s*);
(3) A variable has a size that depends on its type, used CPU and compiler (e.g., size of *s* is 8);
(4) A variable has content which is stored in a specific block of memory assigned by compiler and linker (e.g., content of *str* is the string given by numbers 8 and 3);
(5) A variable has an address pointing to the above block of memory (e.g., *str* is located at address $0x00f3fdf8$) – *that is important for pointer discussion coming next.*

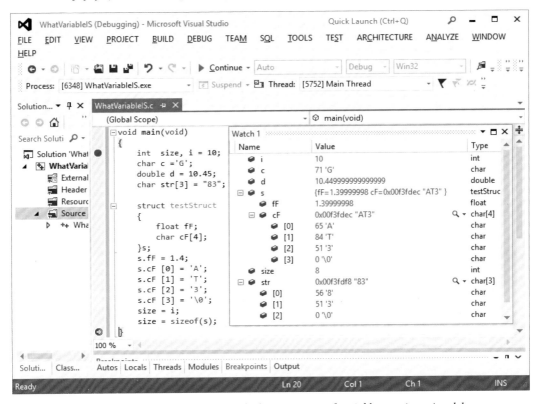

Figure 1.67 What is a variable? A simple demonstration of variable meaning using debugger

As you can see there are two values associated to a variable, its content and its memory location or address. The former is usually designed by *lvalue* (e.g., in "$i = 10;$", variable *i* is a *lvalue*) and the latter by *rvalue* (e.g., in "$size = i;$", variable *i* is a *rvalue*), according to the permitted value relative to the position on an assignment statement. So, *what about adding the following statement "8 = size;" to the above program? Can you figure out the kind of error you will get back? Please try it by yourself.*

Now that it becomes clear why a programmer may need a pointer variable to hold an *lvalue* or an address (i.e., in case there is a need to change it at runtime), let's see what does C language offers to deal with pointers: two special operators '*' and '&' (a.k.a., address operator) to create a pointer and determine the address of the pointer, respectively. The '*' operator is also used to read the value at a given address. Figure 1.68 shows in the first statement of the body of *main()* a pointer variable declaration "*int *pVar;*", instructing the compiler to reserve enough bytes to store an address in memory and that such address is intended to be used to store addresses of integers.

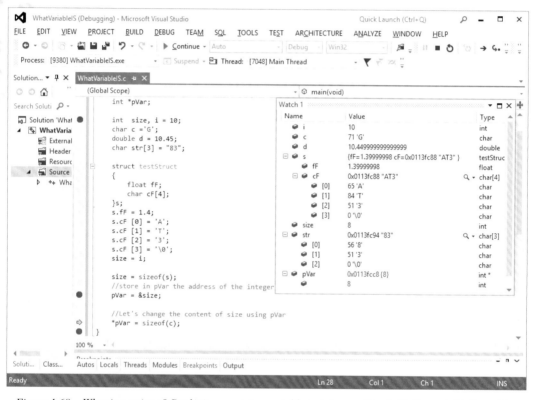

Figure 1.68   What is a pointer? Declaring a pointer variable to integer, pVar, initially not initialize with a specific address as reported by the unknown address

As with any regular variable, let's initialize *pVar* with a known address (i.e., the address of variable *size* as given by the operator '&') while checking if the states of *size* and *pVar* are now the same. Now change the content of the memory block pointed by *pVar* (i.e., using the operator '*') with the size of variable *c*, while also checking whether variables *size* and *pVar* are still in sync in terms of content, as shown in Figure 1.69.

Finally, add a new statement to display the address of the variable *size* to confirm that is exactly the address stored in the pointer variable *pVar* (see Figure 1.70). The realized experiments on pointers still show variable pointers as sharing the same 5 above attributes, but with their contents representing another completely different meaning. They are addresses or references to another program data object, instead of being regular *rvalues*.

1 PROGRAMMING IN C: AN OVERVIEW • 51

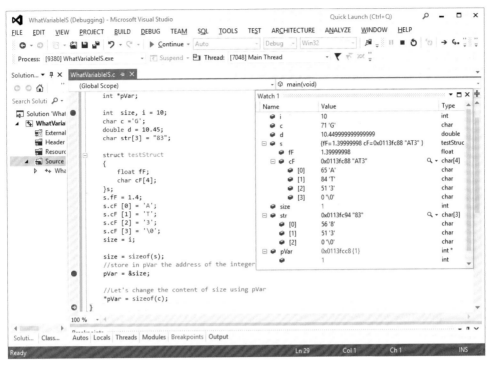

Figure 1.69 Checking through debugging how variables size and pVar are in sync in terms of content as pVar is an alias of variable size

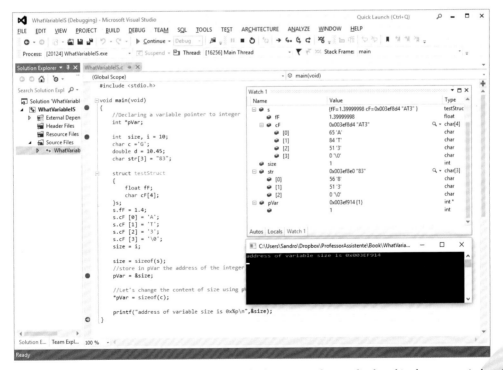

Figure 1.70 Checking that the address in pVar is exactly the same as the one displayed in the output window (i.e., the address of variable size)

Summing up, this chapter shortly guided you to the what of C programming and it will end now with the following recommendations:

Recommendation 0: Never create a projects' folder on the desktop to avoid errors emitted by IDE tools due to the presence of special characters on path names.

Recommendation 1: Always follow C Program Development Cycle, mainly for larger programs.

Recommendation 2: All flowchart graphical symbols, except the decision block should always have only one entry and one exit, both on horizontal sides. Only the decision symbol has two exits, one for the condition testing true and another for testing false.

Recommendation 3: Always initialize any variable, pointer-typed or not, before using or instantiating it. In so doing, you will avoid possible runtime error occurrences due to missing semantic checking.

Recommendation 4: Always assign meaningful names to your program' symbols (e.g., variable and functions) to make the program more readable and consequently more understandable by you and other people as well (i.e., open the door for easy future problem solution enhancement or evolution).

Recommendation 5: Avoid excessive usage of global variables, whenever possible, by implementing functions which through theirs associated local variables limit propagation of malfunction side effects to other parts of the program (e.g., such malfunction can become hard to pinpoint).

Recommendation 6: Always provide function declarations at the top of a source file which invokes them, even if they are all in the same source file.

Recommendation 7: Always name constants using uppercase letter to be easily identified, making the program code more readable.

Recommendation 8: As a naming convention, always postfix with underscore t (i.e. _t), the declaration of a new structure type using *typedef*.

Recommendation 9: Do not ask the instructor or your peers for help before hardly trying yourself by always and firstly "asking" the programmer best friend — the debugger. In so doing, you will enhance your programming skill and autonomy toward instructors and classmates.

The content of this chapter was communicated in a way to leverage learners' attitude and spirit (e.g., intellectual curiosity) towards lifelong learning process through learning by doing, as well as by discovering and exploring programming knowledge by themselves, supported by a programming toolchain. It followed the breadth to the learning approach by briefly presenting "*something of every concepts of the C language*" (i.e., approaching a horizontal teaching).

To attenuate the breadth and depth problem, next chapters will follow the depth to the learning approach by discussing many more details of each previously presented concept and simultaneously presenting real and more complex examples. In general, they will approach a vertical exemplification of teaching to facilitate the development of programming skills with knowledge provided on demand (i.e., on-the-fly) as dictated by the specificity of the problem under analysis.

References

[1] SCHILDT HERBERT. C: The Complete Reference. 4th ed. Berkeley California: McGraw-Hill, 2000.

[2] HOLENDERSKI MIKE. A very brief introduction to programming in C. Eindhoven University of Technology, 2014.

[3] JENSEN TED. A Tutorial on Pointers and Arrays in C, 2003.

Several other internet sources were also used, but they are too much to be referenced here and so, our credits also go towards their authors.

2 HANDS-ON FUNCTIONS

Learning objectives

1. Understanding what is a function and its general skeleton.
2. Understanding function scope and the lifeness of its local variables.
3. Understanding and differentiating function call types.
4. Coding regular functions.
5. Understanding and coding recursion.
6. Practicing problem-solving through algorithm design.
7. Analyzing problems and splitting their solution into several functions.

Theoretical contents

1. Algorithm and programming.
2. Function declaration.
3. Function definition.
4. Function formal parameters, arguments and call types.
5. Scope of a function.
6. Recursive functions.
7. Problems to exercise above contents.

Strategies and activities

1. Always presenting problem statements that must be first deeply analyzed and then follow the remaining phases of C program development cycle. It will be fundamental the choice of varied problems to on-the-fly guarantee, as well as trigger the full coverage of above contents.
2. Always enforcing quality of the developed algorithms, both functionally and graphically.
3. Always promoting modular and reusable designs for future problems and chapters.
4. Always enforcing code quality in terms of named symbols and the way it is commented.
5. Always enforcing at the design phase a full debugging coverage of a function by designing detailed test cases, mainly covering the corner cases.
6. Always forcing deep and isolated debugging of function code before function usage in the final problem solution (i.e., program) based on above test cases.
7. Running a simple quiz about applied knowledge during the design of the problem solution.
8. Ending with a short presentation and discussion of at least two problems (i.e., a good and a bad solution) and their solutions.

C is a structured language which uses global function as its main structural component to leverage compartmentalization of code and data. Basically, a function represents the building block which encapsulate all data and statements' sequence of a specific task from all the remained program code. Usually, a function should represent a conceptual unit or task about which it is possible and useful to think and reason of in isolation (i.e., a self-contained block of code), and so, allowing larger projects to be programmed in a modular way. However, identifying candidate functions and their behaviors in a program precedes programming itself and it comes under the problem-solving process of finding solutions to problems (i.e., it is carried out at the first three steps of the C program development life cycle). When approaching a problem, all variables and their associated requirements and constraints will be enumerated at the analysis phase, while algorithms for candidate functions are developed at the design phase. In general, an algorithm can be defined as a step-by-step procedure to be taken for producing a specific result and it can be expressed using different notations, such as pseudocode and flowchart. While algorithm is a generic method, a program in C is a specific method or an instance of the algorithm that has been customized to solve a specific task using C language notations. Thus, programming can be described as the process of encoding an algorithm into a notation of some programming language (i.e., as done at the implementation phase of C program development life cycle).

2.1 Algorithm and Programming

There is no real programming as an activity and a program as its output, at least for important problems requiring reasonable intellectual effort, without an algorithm representing the problem solution. Therefore, the important first step is the need to find the problem solution or algorithm by carrying out a problem-solving, which can be simply defined as a 3-step generic or ad hoc method of finding solutions to complex problems or achieving objectives. It can be directly mapped to "problem statement", analysis and design phases of C program development life cycle as described in Chapter 1, section 1, and represented by Figure 1.1. After knowing the real problem, major decision points, inputs, outputs and their associated constraints should be identified and only then designing the algorithm which is the problem-solving output. Designing the algorithm is per se a multiple-step process, performed by (1) discovering a general algorithm, (2) generating some alternative algorithms, (3) successively refining algorithms to details close to the chosen programming language, and (4) evaluating and selecting alternatives. Figure 2.1 will shortly presents you with a view of what a problem-solving is and its relationship with problem, algorithm and program. It is obvious that for each problem or class of problems, there may be many different algorithms, while for each algorithm, there may be many different implementations (i.e., programs which are instances of an algorithm, possibly coded by different programmers in different languages). As an ad hoc method, problem-solving will require time and practice to do the design well, while developing functional skills to break the problem solution down into smaller different problems' solutions (i.e., according to the rationale of logically splitting the original solution into single-functionality tasks). In so doing, the programmer will not be overwhelmed by the entire and more complex problem as it can be approached by a series of smaller solvable problems, i.e., into a more manageable size.

Although problem-solving was described above as a 3-step method, some literature also defines it as a 4-step method, including the implementation phase (i.e., the programming process). No matter the number of steps (3 or 4) of a problem-solving approach in the context of computer programming, its final output is always an algorithm — a generic one or a program that is always an instance or a specialization of the generic algorithm produced at the design phase.

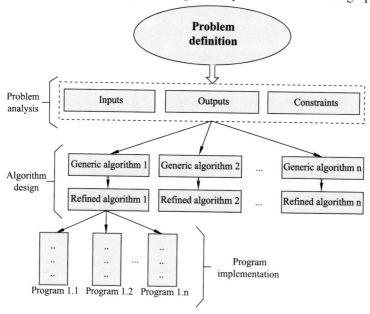

Figure 2.1 Presentation of problem-solving as a 4-step process in programming context

For an introductory programming learning environment, like the one described in this book, the 4-step approach for the problem-solving will fit well because the size and complexity of problems at hand are much smaller and easier compared to industrial or business environments. In latter environments, bigger project team is involved, usually consisting of more than one person. However, throughout this book the more generic 3-step approach will be used and extended with the "implementation and testing phase" to define the programming process. In doing so, we excluded the evolution phase from the programming process, mainly due to the size of the problems at hand.

Algorithm is used in computing and also in completely different domains such as mathematics, linguistics or even in our daily life. It has been formally defined in the literature as a sequence of well-ordered operations for completing a task by giving an initial state, proceeding through well-defined series of successive states, and eventually terminating in an end-state. Examples of algorithms are: a cooking recipe, making a diagnosis, or a probabilistic algorithm. Main reasons such as efficiency, abstraction and reusability have been pointed for the use of algorithms, which can be expressed in many kinds of notations (e.g., programming language, flowchart and pseudocode). Since the focus goes towards teaching programming algorithm to newbies, the following definition for algorithm will be used throughout this book: *"a recipe describing the exact sequence of steps needed to be performed by the computer to solve a problem or reach a goal"*. Generically, the following properties have been recognized in a good algorithm:

1. *Unambiguous and well-ordered* since the steps should be precisely stated or defined and in a clear order;
2. *Specified input* as data to be transformed during the computation to produce results along with its type, amount and data format;
3. *Specified output* as data resulting from the computation alongside its intended result or clearly specifying that no output is produced;
4. *Definiteness* as the specification of the sequence of events, describing details of each step, including how to handle errors;
5. *Finiteness* meaning the algorithm must eventually terminate after the execution of a number of instructions;
6. *Effectiveness* meaning all operations to be performed in the algorithm are sufficiently basic to be doable in a finite length of time.

Also recognized are the following generic or common elements of algorithms of which some are easily mapped to structured language constructs such as *while*, *do⋯while* or *for*:

1. *Processing* as means to perform arithmetic computations, comparisons, and testing logical conditions;
2. *Acquiring* data as means to read specific problem data values from an external source;
3. *Reporting results as* means to inform users about computed results;
4. *Selection as means* to choose among possible execution paths according to initial data, user input and/or computed results;
5. *Iteration* as means of repeatedly executing a sequence of operation, for a fixed number of times or until some logical condition holds.

Flowchart notation was chosen to express and visualize algorithms throughout this book because of its (1) unambiguous representation compared to natural language, (2) independence from implementation programming language, (3) ease of communication to others, (4) effective and efficient analysis as well as coding, and (5) proper documentation and maintenance. However, it has also inherent drawbacks in supporting:

1. *The representation of complex problems* because flowcharting can become very complicate and difficult to handle;
2. *Multiple modifications* as any change to a complex flowchart may demand a complete re-

drawing.

Flowchart has been defined as a diagrammatic representation which illustrates a solution model (e.g., the steps and structure of an algorithm or program) to a given problem. Basically, it shows the flow of control of an algorithm as it runs step-by-step in a visual manner and so, it helps understanding process, as well as finding flaws and bottlenecks on it. Table 2.1 presents the most common boxes and notational conventions of flowchart such as the *activity* (i.e., processing step) and *decision* denoted by a rectangular and diamond boxes, respectively.

Table 2.1   A repertoire of boxes and notational conventions of flowchart

Symbol	Name	Meaning
▭	Process	Represented as a rectangle, it indicates any type of internal operation inside the processor or memory: data transformation, data movement, logic operation, etc
▢	Terminal	Represented as a rounded (fillet) rectangle, it indicates the starting or ending of the program or algorithm
◇	Decision	Represented as a diamond, it evaluates a condition or statement and branches depending on whether the evaluation is true or false
▱	Input/Output	Represented as a parallelogram, it involves receiving data and displaying processed data
▯	Predefined Process	Represented as a rectangle with double-struck vertical edges, it indicates a subroutine or complex processing steps which may be detailed in a separate flowchart
○	Connector	Represented with a circle, it indicates where multiple control flows converge in a single exit flow
⇄ ↑↓	Flow Lines	Represented with arrows, it indicates that control passes from one symbol to another symbol in the direction the arrow points to

Figure 2.2 depicts a generic flowchart illustrating the use of the most common symbols. The flowchart illustrates a simple input-decision-process-output program. The program starts by gathering some data from a specific input channel. Then, the flow is driven by a block decision: if the condition is false, the program requests a new input; otherwise, the data is then processed and the result is finally outputted/displayed.

By previously defining problem-solving and consequently the algorithm that is at its core as an ad hoc method, it becomes clear that designing algorithms will be learned only by experience: first, by seeing others solving problems and then by doing it yourself. Furthermore, flowcharting is a highly iterative process, leading you to successively refining algorithms to get it right. Throughout this book you will be exposed to a considerable number of different algorithm designs, pushing you to develop pattern recognition and then apply those patterns later to solve similar problems.

Finally, to theoretically conclude this paragraph some generic recommendations or rules for flowcharting are enumerated:

1. All symbols of the flowchart are connected with arrows (not lines);
2. Flowcharts are drawn so flow generally goes from top to bottom;

3. All flowchart graphical symbols, except the decision block should always have only one entry and one exit on top and bottom, respectively.
4. The decision symbol has two exits (one from the bottom and another from the sides) for the condition testing true and another for testing false;
5. Connectors are used to connect breaks in the flowchart, e.g., from one page to another;
6. Each subroutine or function program has its own and independent flowchart;
7. The beginning and the end of a flowchart is indicated using the terminal symbol, unless it ends with a contentious loop.

Figure 2.3, Figure2.4 and Figure 2.5 show alternatives algorithms to the ones presented in Figure 1.2 and Figure 1.65, using not only a pseudocode representation but also another flowchart and a programming algorithm in C.

Figure 2.2 Generic flowchart illustrating the use of the most common symbols

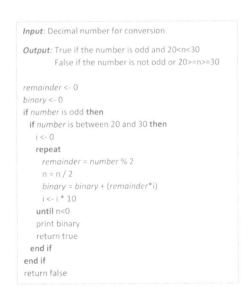

Figure 2.3 A pseudocode algorithm expressing the decimal2binary conversion for odd numbers greater than 20 and lower than 30

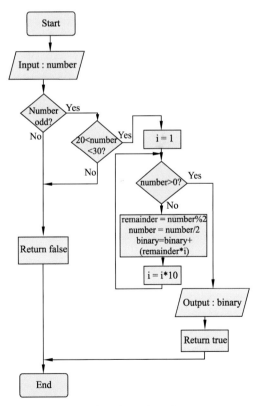

Figure 2.4 A flowchart algorithm expressing the decimal2binary conversion for odd numbers greater than 20 and lower than 30

To conclude this paragraph, let's briefly demonstrate the use of the connector symbols, as well as how to leverage flowcharting modularity. *Imagine you want to read and convert in sequence two odd decimal numbers greater than 20 and lower then 30 to binary and display their sequence of bits on the screen"*. The first solution could be a big flowchart as represented in Figure 2.6, where two connector symbols were used to connect separated parts for reading, checking, converting and displaying each number. Such solution was too big mainly because it replicated exactly the same sequence of steps twice for each read number. A more modular solution could be presented by encapsulation the replicated part in only one predefined process as represented in Figure 2.4 and then using it to process the two numbers (see Figure 2.7).

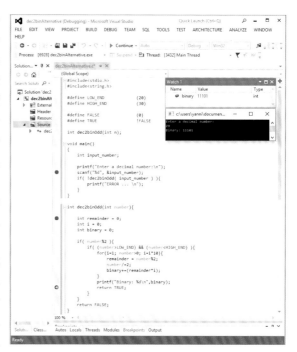

Figure 2.5 A programming algorithm expressing in C notation a specialization of the flowchart algorithm for decimal2binary conversion of odd numbers greater than 20 and lower than 30

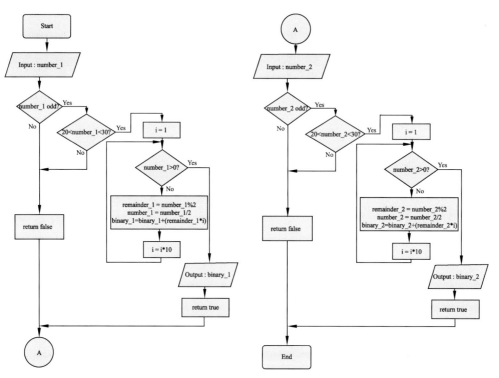

Figure 2.6 The use of flowchart on-page connector symbol to represent a large algorithm with two parts on the same page. It reads and converts two odd decimal numbers to their binary representation

What if the two parts of the algorithm in Figure 2.6 are so big and they do not fit on the same page? Therefore, the on-page connectors should be changed with off-page connectors accordingly labeled to indicate the pages where processes (i.e., algorithm parts) should be continued, as shown in Figure 2.8. Basically, on-page and off-page connectors should be used to improve the readability of a flowchart by avoiding crossed lines on a page and, overcrowding as well as moving action between pages, respectively.

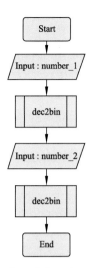

Figure 2.7   A modular algorithm to read and convert two odd decimal numbers to their binary representation by using the algorithm in Figure 2.4

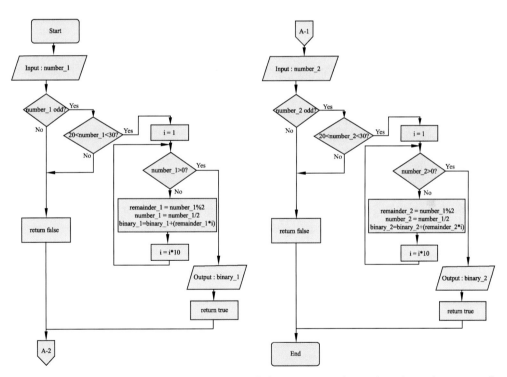

Figure 2.8   The use of flowchart off-page connector symbol to represent a large algorithm with two parts flowing on different pages

2.2 Function Declaration and Definition

A function in C is nothing more than a programming algorithm which separates and packages sequence of statements (i.e., the body of the function which also includes the return statement, as well as some local declarations) by enclosing it between the function declaration or header and the closing bracket. Function declaration, prototype and definition were briefly introduced in Listing 1.1 of Chapter 1, while now they will be deeply reviewed, starting with a generic structure of C language function as presented on Table 2.2. Basically, a function definition has two parts: the function declaration and the body.

As mentioned in previous Chapter 1, the function declaration provides not only to the compiler but also to the caller of a function with the basic information regarding the usage of the callee function by specifying how actual parameters (i.e., arguments) and generated results are passed to the callee function and transmitted back to the caller function, respectively. The function declaration is the function interface to the outside world and it offers the following information to the compiler and the caller function:

1. *The function return type*: not all functions need to return a generated value and in this case the return type must be *void* (e.g., as in Figure 1.65). The return statement terminates the callee execution and returns control back to the caller at the statement immediately following the call. Optionally, its syntax allows for a return value and if present, the value of the expression is evaluated and automatically converted to the return type of the function as specified on its prototype.

2. *The name of the function*: the actual name of the function is a regular user-defined C token which is used by the caller function to invoke the callee function. Whenever possible, function names should always start with a verb.

3. *The list of formal parameters*: it is optional as a function may contain no parameters and basically, it refers to the type, order, and number of the parameters of a function. A parameter is a local storage to a function which is used to pass a value. Saying it in another words, the value of the actual parameter or argument specified in a call of a function is assigned to the parameter.

Table 2.2 The general form of function definition in C programming language

Definitions	C Program	Comments
The function declaration	int dec2binOdd(int number)	Indicating the beginning of a function definition, the return type, the function name and the list of formal parameter (i.e., in this case only an integer variable number)
The opening bracket	{	Indicating the start of the function body

The function body	int remainder = 0;
	int i = 0;
	int binary = 0;
		Local declarations which along with the formal parameters leverage the compartmentalization and so, the independence of the local memory

Definitions	C Program	Comments

The function body	if(number%2) {
	  if((number>LOW_END) && (number<HIGH_END)){
	  for(i=1; number>0; i=i*10){
	    remainder = number%2;
	    number/=2;
	    binary+=(remainder*i);
	  }
	  printf("Binary: %d\n",binary);
	  return TRUE;
	  }
	}
	return FALSE;
		The sequence of statements which define the processing steps to be performed by the function

The closing bracket	}	Indicating the end of the function body and so, the end of function definition. This point is reached during the execution, if no return statement appears in a function definition and the control automatically returns to the caller function

Let's coding a C program that is able to calculate an approximate value of the square root of a number, while playing a little with the return statement. Starting with the analysis, let's constrain the solution space to the set of real numbers, and process only positive real numbers as the square root of a negative number belongs to the set of complex number. The square root of a positive real number can be approximately calculated using Equation (2-1) (i.e., Newton–Raphson formula), shown below, under a controlled accuracy maximum number of iterations.

$$\begin{aligned} s_k+1 &= (s_k+x/s_k)/2 & k \geqslant 1 \\ s_1 &= x/2 & x > 1 \\ s_0 &= x \end{aligned} \quad (2\text{-}1)$$

Having already the data specification for the problem (i.e., a positive real number to be processed and a constant specifying the accuracy), the problem solution can be conceived using the following 3-stage pattern of getting data, processing data and presenting results. Let's then, iteratively refine the algorithm as shown in Figure 2.9, by flowcharting from a coarse-grained to middle-grained and finally to fine-grained with enough detail to be straightly converted to a C programming algorithm.

As an exercise, we shall ask you to encode the flowchart on Figure 2.9 c) to C language notation, while we shall redraw it to a more modular solution by calling a function (see Figure 2.10).

Let's now move to the implementation stage by coding both algorithms and then debugging them to check if the final program is working accordingly. Note that you can also call the pre-defined *sqrt*() provided by C standard library for mathematical computations, *libm*, and prototyped at <math.h>. Table 2.3 presents a dissection of the C programming algorithm for the generic algorithms presented in Figure 2.10.

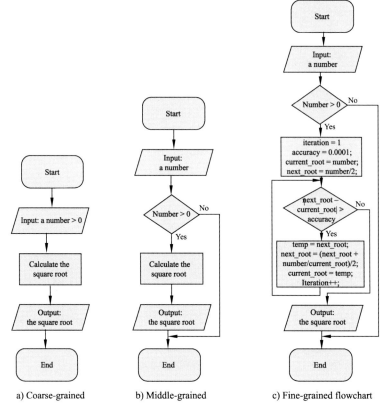

Figure 2.9   Multi-level flowcharting

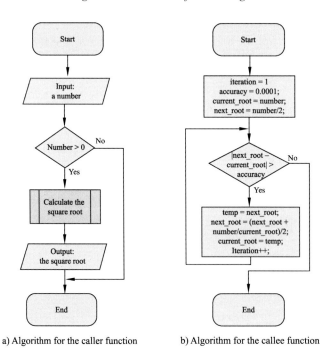

Figure 2.10   A modular alternative to the problem solution presented on Figure 2.9 c)

So far two main aspects of a function had been tackled, i.e., function declaration and function definition, but there is still a third crucial aspect, regarding the means for using a function, also known as function application, or function call. To use a function, you will have to call it to perform the defined task and such call can be made through its name while passing the list of arguments. In Figure 1.54 in Chapter 1 the function call construct was slightly presented and discussed without details about its associated syntax:

function_name (argument$_1$, argument$_2$, $\cdots$, argumentn);

As you know so far, not all functions take arguments, but for those functions defined with a non-empty list of parameters, the list of arguments which are passed for the call should agree in type, number, and order with the list of parameters in the function declaration. *What happens if an argument to a function does not match the type of the corresponding parameter?* Remember that C language leaves to the programmer the duty for checking the type of compatibility between a parameter and an argument during the call of a function. It will automatically convert argument values to parameter types where possible (e.g., sometimes with dangerous side-effects) and otherwise, some runtime errors will be later triggered (see Figure 1.36 in Chapter 1).

Table 2.3 Dissecting line by line the Dec2bin program

C Program	Comments
#include <stdio.h>	*stdio.h* is the standard header file with the function prototype for scanf().
#include <math.h>	*math.h* is the standard header file with the function prototype for fabs().
double SqrtUsingNewtonRaphson(double positiveNumber, double approximationAccuracy); //double SqrtUsingNewtonRaphson(double, double);	Function declaration or prototype for *SqrtUsingNewtonRaphson()* with long and meaningful names of parameters to better understanding the meaning of each parameter. Alternatively, a prototype can omit the name and meaning of parameters but is not recommended.
void main(void)	Starts the definition of the *main()* by including its declaration as a function that returns nothing and also receives no argument
{	Begins the body of *main()* with the left bracket symbol.
double root;	Declares a real number local variable named root.
scanf("%lf",&root);	*scanf()* is waiting for an real number to be entered by the user and stores it in the location assigned to the variable named root.
if(root > 0)	Check if the read number on root is positive by comparing it with zero
root = SqrtUsingNewtonRaphson(root, 0.0001);	Calculate the square root only if the read number is positive by invoking *SqrtUsingNewtonRaphson()*. The function was called by passing the arguments or actual parameters 0.0001 and the read real positive number.
}	Ends the body of *main()* with the right bracket symbol.
double SqrtUsingNewtonRaphson(double number, double accuracy)	Starts the definition of *SqrtUsingNewtonRaphson()* by its declaration as a function returning a double value and taking two double numbers as inputs.
{	Begins the body of *SqrtUsingNewtonRaphson()* with the left bracket symbol.
int iteration = 1;	Declares an integer local variable iteration and initialize it outside the loop body to indicate the first iteration performed.
double temp,current_root = number;	Declare 2 double variables *temp and current_root* and initialize the latter with the value for the *S0* of Newton Raphson formula.
double next_root = number / 2;	Declare another double variable *next_root* and initialize it with the value for the *S1* of Newton Raphson formula.

Continued

C Program	Comments
while(fabs(next_root – current_root) > accuracy)	Checks if good accuracy or acceptable error was achieved in the last performed iteration, calling *fabs*() for the absolute value of the error.
{	Opens the multiple statements while-block with left bracket symbol.
temp = next_root;	If desired accuracy is not achieved yet, previous Sk+1 is temporarily memorized for later used as Sk.
next_root = (next_root + number/current_root)/2;	Calculate the current value for Sk+1.
current_root = temp;	Update accordingly Sk with the previous calculate value of Sk+1.
iteration++;	Update the number of iteration performed so far. Notice that this value is not used by now but we shall come later to it.
} return next_root;	Closes the multiple statements *while* block with right bracket symbol and return to the caller along with calculated root square of the input number
}	Ends the body of *SqrtUsingNewtonRaphson()* with the right bracket symbol.

Let's play a little with the above program algorithms by creating a MS Visual Studio project and then setting the shown breakpoints while running in debugging mode to first illustrate the passing of parameters from the caller *main*() and then the return from the callee *SqrtUsingNewtonRaphson*(). Figure 2.11 is the debugging view after pressing F5 twice, taking you to the start of the function body. Looking at the *Watch window*, arguments 24.6 and 0.0001 were effectively passed to *SqrtUsingNewtonRaphson*() through the local storages assigned to the formal parameters presented in the function declaration, i.e., through parameters *number* and *accuracy*, respectively. Pressing F5 once again, the sequence of statements in the body of *SqrtUsingNewtonRaphson*() is executed and you'll see on the *Watch window* of Figure 2.12, the calculated values of all local variables, e.g., 25 as the number of iteration needed to achieve the required accuracy for the approximated square root as 24.6.

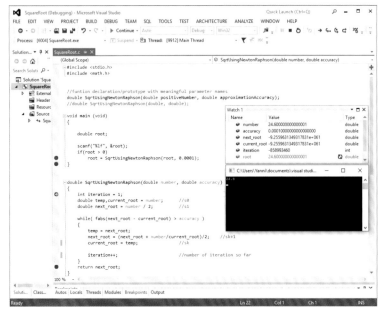

Figure 2.11   Watch window shows the arguments 24.6 and 0.0001 as passed during the function call through the local storages assigned to the formal parameters number and accuracy, respectively

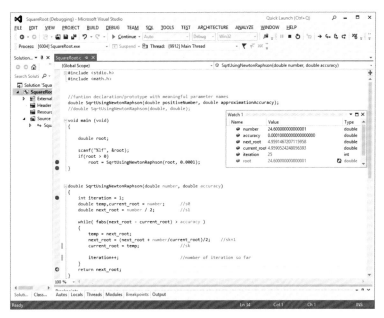

Figure 2.12   Content of local variables to the callee SqrtUsingNewtonRaphson(), including the two parameters before returning to the caller main() (i.e., while executing inside the body of callee)

Pressing F5 once again (see Figure 2.13), the control is returned to the caller which is the *main*() after the call statement, and all local variables to *SqrtUsingNewtonRaphson*() will be no more in a valid state as they became out of scope (i.e., they are not accessed by or available to the caller).

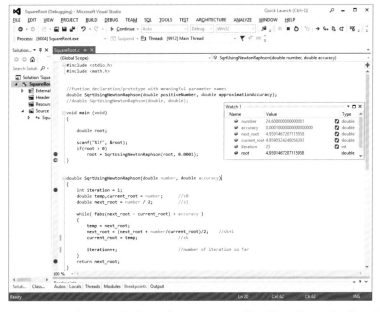

Figure 2.13   Control is returned to main() after ending execution of SqrtUsingNewtonRaphson() at the left bracket which end the body of main() as there is no statement after the call statement. The lifetime of all local variables local to SqrtUsingNewtonRaphson() is over as they are out of the scope of main()

2.3 Function Call Types and Their Mechanisms

Looking to Figure 2.14 and more specifically at the two call sites in *main*() altogether with the two companion descriptions in step 1 and step 3, it becomes obvious that there is some missing detail regarding to the function call construct.

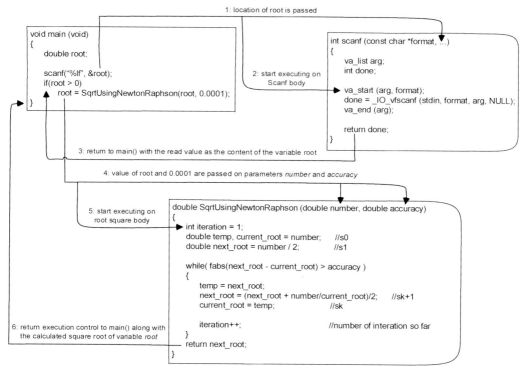

Figure 2.14 Execution flow of the program presented on Figure 2.13. The flow to fabs() was omitted because it could be implemented as a macro, an assembly routine or in C but using inline assembly. By now it was interpreted as a macro to not overload a novice programmer with too much details or low-level stuff

Let's play a little with function call while trying to figure out what is the missing detail by answering: *what if you change the value of variable number in the callee body?* Let's change a little bit the previous program by adding the following statement "*number /= next root;*" and also declaring a new variable to separate the read number from its approximate square root. Then let's enter the debugging mode and press F5 twice to trigger the second call site (see Figure 2.15).

Since the values or contents of argument *num* and parameter *number* are different at the *Watch* window, one can conclude that they do not share the same local storage, i.e., the argument *num* was directly copied into a different storage location which was assigned to the formal parameter *number*. Pressing once again F5 to return execution back to *main*(), both *num* and *number* still keep different values, with *num* preserving its original value (see Figure 2.16), in spite of the its modification in the body of *SqrtUsingNewtonRaphson*(). This parameter passing mechanism is named call-by-value or pass-by-value and it can be described by the following characteristics:

1. A duplicate copy of the argument is passed to the callee through the formal parameter;

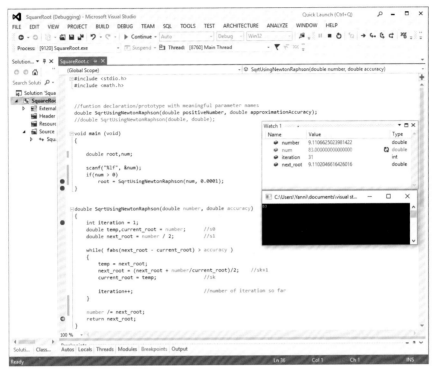

Figure 2.15 Trying to figure out call-by-value parameter passing details at the second call site as commented in step2 in Figure 2.14

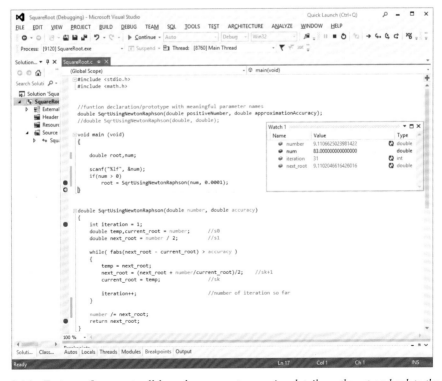

Figure 2.16 Trying to figure out call-by-value parameter passing details on the return back to the caller

2. Modification to the formal parameter in the callee body has no effect on the argument on the caller.

Let's keep playing with parameter passing mechanism while trying to figure out more missing detail by answering: *what about if you want to return back to the caller the number of iterations needed to achieve the desired accuracy?* Since the return statement can only communicate a single-value back to the caller, then let's change the declaration of *SqrtUsingNewtonRaphson()* to accommodate another formal parameter as a pointer to integer (see Figure 2.17). It's the same mechanism used at the first call site to invoke *scanf()*.

Repeating the previous debugging steps while looking at the *Watch window* in Figure 2.17, it is visible that both the argument (i.e., *iter*) and the formal parameter (i.e., *iteration*) share the same local storage at address 0x010FF794, although they do not show the same content only because the argument, *iter*, is not directly accessible from *SqrtUsingNewtonRaphson()*. Then, pressing F5 to return execution back to *main()*, you will realize that the argument, *iter*, was accordingly updated with the same value as the formal parameter, *iteration* (see Figure 2.18). In other words, the callee *SqrtUsingNewtonRaphson()* modified the caller *main()* memory. This parameter passing mechanism is named call-by-reference or pass-by-reference and it can be described by the following characteristics:

1. The actual copy of the argument is passed to the callee by copying its address to the formal parameter, i.e., both argument and parameter will share the same local storage;
2. Modification to the formal parameter in the callee body will affect the value of the argument on the caller as both share the same local storage.

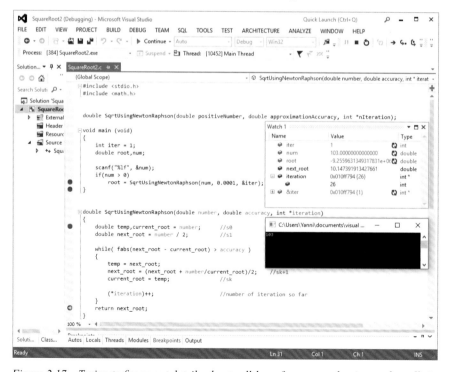

Figure 2.17   Trying to figure out details about call-by-reference mechanism at the call site

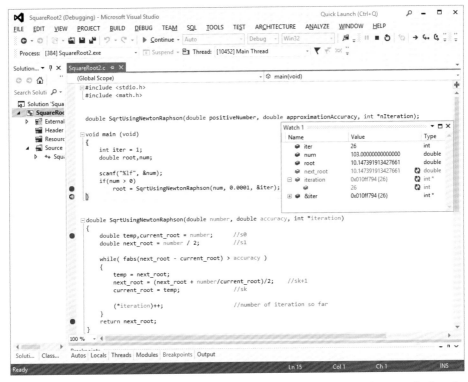

Figure 2.18   Trying to figure out call-by-reference parameter passing details on the return back to the caller

As you can see from the program on Figure 2.18, the C language does not support reference parameters automatically and so, the programmer must implement it manually using pointer constructs. At the call site, the caller should pass a pointer to the callee (e.g., using the & operator to calculate the address of the argument as done with *&iter*), while at the callee body that pointer needed to be dereferenced using the operator * as in the statement *(*iteration)++;*. *Which parameter passing mechanism to choose?* It depends on a case-by-case analysis while tackling different parts of the program under implementation, and the following points will help you to wisely decide for the best tradeoff:

1. *Expressions can be passed as an argument only by value*, because call-by-reference constrains the argument to refer to a specific instance of the formal parameter type stored in programmer-accessible memory (see Figure 2.19 for the argument "*2*num*");
2. *Call-by-reference alleviates the return statement limitation* of only communicating a single-value back to the caller (see Figure 2.17 with a callee communicating to the caller not only the approximated root, but also the number of iteration);
3. *Call-by-reference breaks a little the compartmentalization of data* promoted by C language through the independence of the local memory (see Figure 2.18 where changing a parameter in the callee will propagate to the caller, if wrongly done);
4. *Duplicating a copy of an argument as done with call-by-value can be expensive* in terms of performance, if the argument content is very large. This point will be later discussed and explored on coming chapters.

Figure 2.19 also shows that a function with a return type different of *void* can be part of an expression in C, e.g., *"root = 2*SqrtUsingNewtonRaphson(2*num, 0.0001, &iter);"*.

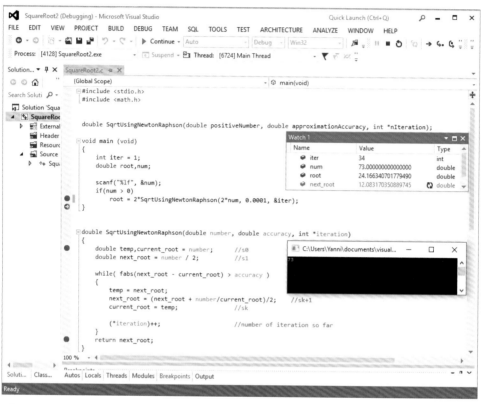

Figure 2.19 Illustrating expression passing as an argument by-value and a function that does not return void directly used on an expression

2.4 Function and Programming Modularity

So far you have been exposed to several function aspects as well as to some benefits in using function, like a better understanding of a program. Such better understanding is a direct consequence of compartmentalization of code and data and it is only achieved if functions are not overused on the program. Therefore the points are: *Are there any more advantages that can be leveraged by the use of functions? Which are the main disadvantages in using functions?*

Rephrasing this chapter introduction, the use of functions is a programming style promoting reusable code that can be called whenever required, instead of replicating the same code in multiple places in the program (please recall to algorithms shown in Figure 2.6 and 2.7). Function on programming languages supports the modularization process, a divide and conquer algorithm design paradigm, carried out during the problem solving, to break down large and complex problems into easily solvable, manageable, and functional units. By coding these functional units as compartmentalized or loosely-coupled C functions with their body insulated from each another, one only needs to know what each function does, instead of how they provide or perform their related functionalities.

Trying to answer the two above questions, we should also point that as important as splitting a complex problem into independent building blocks or modules, the right tradeoff between the number and size of such modules should be guaranteed to avoid an overuse of functions on the final program, because it can easily nullify the gained readability and understanding of a program. Summing up, let's enumerate other advantages of using functions such as:

1. Functions are made for code reusability as well as for reducing coding time and overall program memory footprint as duplicate sequence of statements are replaced by function calls;
2. Implementing functions also simplifies debugging while reducing debugging time due to error isolation and containment to the function body, instead of spreading the same errors throughout the program and then fail to pinpoint and fix some of them (i.e., implementing functions promotes a king of "*find it once and eliminate it quickly*" approach);
3. Implementing functions also improves team collaboration in the development of larger projects, as programmers on the same team or on independent teams can code them separately without accidentally affecting each other's code (i.e., without creating side-effects in other parts of the final program);

Here are also some pointed disadvantages of using functions as a programming style:

1. Using too much fine-grained functions can make the understanding and readability of a program much harder, leading to maintenance problems and therefore, longer time to pinpoint and fix triggered runtime errors or bugs;
2. Function call does not come for free as it eats CPU cycle time. However, such performance penalty should be wisely profiled in a case-by-case basis as excessive inlining of functions (i.e., duplicating functions' body sequence of statements by replacing call sites) can cause non-trivial amounts of code size, leading to performance losses due to modern CPU features (e.g., caching behavior).

For a short illustration of above aspects, let's implement a C program for the following problem: *imagine an array of 18 integer numbers which possibly should represent numerical coefficients of quadratic equations. Determine: (1) the roots for those equations with solution in the real space, (2) the least common multiple of last two coefficients for those with roots on complex space and (3) the binary representation of the second coefficient if the coefficients do not represent a quadratic equation.*

In spite of being a small problem, let's try a modular design to illustrate some of the points discussed above. From the above problem statement, let's move to the analysis phase trying to understand the core of the problem. So, you need to know that a quadratic equation is generically represented by $ax^2 + bx + c = 0$, with the 3 coefficients a, b and c all type of integer as dictated by the problem statement (i.e., recall that the array has 18 integer values which possibly represents 6 equations, i.e., 18/3). One method to find the roots is applying the quadratic formula given by Equation 2-2, but there are other alternatives, e.g., applying

Lagrange's resolvents.

$$x = \frac{-b \pm \sqrt{b^2 - 4ac}}{2a} \qquad (2\text{-}2)$$

The discriminant given by b^2-4ac determines where the roots are located. A negative discriminant indicates roots on complex space. To calculate the square root you can use the algorithms presented on Figure 2.9 b) and Table 2.2 as the problem solution and function definition demands for roots in the real space and so, promoting reusable code. Furthermore, you can use the *sqrt()* provided by *libm* and prototyped on *math.h*, if you want a generic solution with roots on both spaces.

Knowing how to check for the complex space of solution using the discriminant, you need then to know: what is the least common multiple (LCM) of two numbers? It is defined as the smallest number that is a multiple of both of them. The simplest method to find *LCM(b,c)* is by listing multiples of b and c until you identify the smallest multiple they have in common. Alternatively, you can also apply the Euclidean algorithm, to compute the greatest common divisor (GCD) of two usually positive integers and then divide their product to the *GCD(a, b)*, i.e., *LCM(b,c) = (axb)/GDC(b,c)*. Finally, you need to know that to be a quadratic equation, the coefficient *a* should be non-zero (i.e., a≠0). For the conversion from decimal to binary, you can simply promote the algorithms presented in Figure 2.4 and Table 2.1. However, as coefficients on the array can take negative values, argument passed to the programming algorithm on Table 2.1 will be the absolute value of the coefficient *b*.

Moving to the design phase while reviewing the analysis phase ideas and results, at least 4 modules were identified for: (1) solving quadratic equations, (2) converting from decimal to binary, (3) calculating the square root of a number, and (4) finding the least common multiple. Each module is represented by one flowchart algorithm, except LCM module which demands for an algorithm to find the GCD, in case the Euclidean approach is followed. Flowchart algorithms for these modules are presented as a partial problem solution in Figure 2.20 and Figure 2.21, as well as on Figure 2.9 c) and Figure 2.4.

However, to fully represent the problem solution, another *main* module which will integrate all previous ones will be presented on Figure 2.21. Figure 2.22 presents an alternative algorithm for *LCM (b,c)*, i.e., the simplest strategy that lists multiples of b and c until identify the smallest multiple they have in common.

Notice that the way the problem solution was broken down into four modules has only the main purpose to give you some insights about how to approach a modular design for the problem at hand. Of course, that exist many other solutions, at least approaching a more coarse-gained decompositions, e.g., by merging some of mathematics algorithm (e.g., merging LCM with GDC, quadratic solver with square root, and so on), instead of splitting them into several modules as done above. By practicing, you will gain enough functional skills to easily find the right decomposition tradeoff for the problem you have at hand, under some constraints.

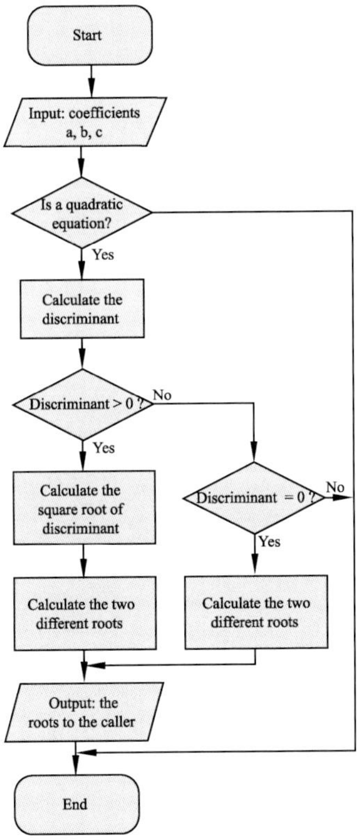

Figure 2.20 Algorithms for the independent modules to calculate a) the quadratic roots, b) the least common multiple and c) the Euclidean greatest common divisor

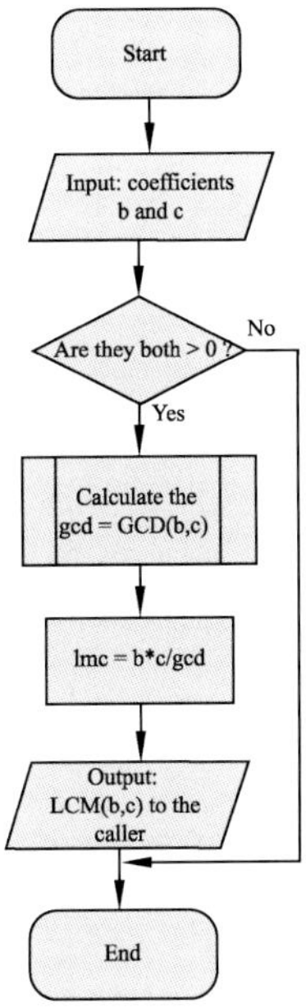

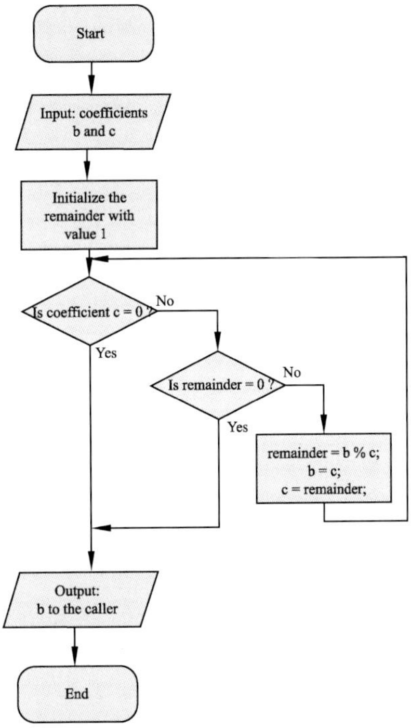

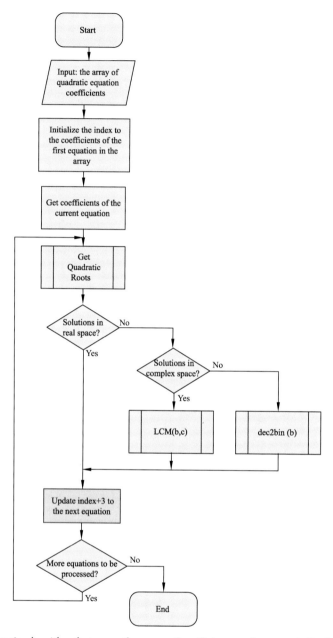

Figure 2.21   The main algorithm that scans the array of coefficients and process individually each equation

Such modularity should also be leveraged at the '*Implementation and Testing*' phase of the programming process. Hence, and also as all functions must be prototyped before use, let's then apply a universally followed convention for C programs which provides multiple header files (i.e., .h files) with functions' prototypes for each modules, instead of prototyping them all at the beginning of the source file (i.e., .c files), as have been done so far throughout this chapter.

The implementation file, *dec2bin.c*, shown in Figure 2.23 with the programming algorithm for the generic algorithm on Figure 2.4, starts with the line *#include "dec2bin.h"* to force the "*prototype before definition*" rule, and so, making the prototype as well as other symbols defined in *dec2bin.h* visible to the definition of *dec2bin()*. Furthermore, any implementation file calling *dec2bin()*, e.g., the main module, *main.c*, in Figure 2.25 must include the line *#include "dec2bin.h"*.

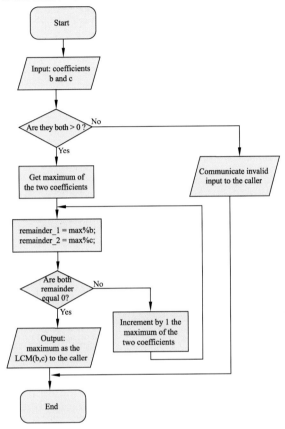

Figure 2.22   Alternative algorithm for LCM(a, c) by listing the multiples of b and c

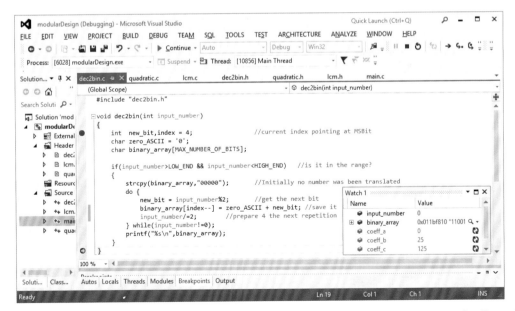

Figure 2.23   Implementation file with the programing algorithm for the main algorithm as presented in Figure 2.4

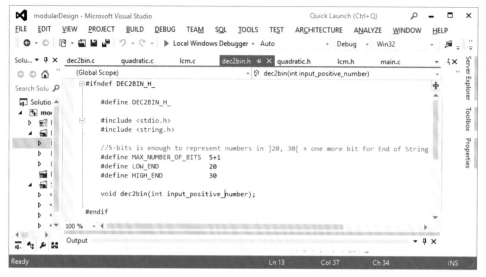

Figure 2.24   Header file defining dec2bin() prototype and other symbols needed by dec2bin() definition

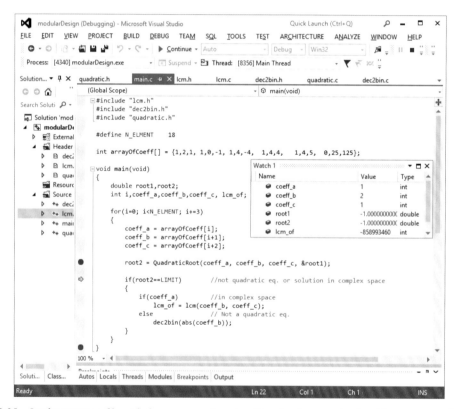

Figure 2.25   Implementation file with the programming algorithm for the generic algorithm converter from decimal to binary presented in Figure 2.21

Looking at header files shown by Figures 2.24, Figure 2.26 and Figure 2.27, it is obvious they all follow the same and the following template based on the header guard construct (a.k.a., *wrapper #ifndef*) described on Table 2.4.

2 HANDS-ON FUNCTIONS 79

Table 2.4 Header guard template to protecte a header file (e.g., filename.h)

C Program	Comments
#ifndef FILENAME_H_	Check if the symbol, *FILENAME_H_*, has been previously defined. This symbol should be unique and in spite of being arbitrary, it is recommended to always use the name of the file with some additional text – normally, filenames use to be unique in a given project.
#define FILENAME_H_	Define the unique symbol as stated above. This unique symbol is named *controlling macro* or *guard macro*.
/*insert here your code*/	You can insert real code, but usually it contains C declarations and macro definitions which will be included only once per each translation unit.
#endif	

Figure 2.26 Header file defining function prototypes and other symbols supporting the definition of sqrt() and QuadraticRoot()

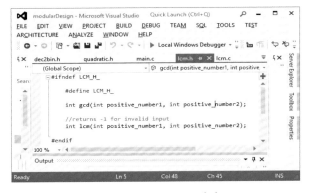

Figure 2.27 Header file defining gcd() and lcm() prototypes needed to support prototype before use of by gcd() and lcm() definitions

Header guard is a preprocessor technique to prevent a shared header file between several .c files from being included multiple times by checking whether a *guard macro* (e.g., FILENAME_H_) has been defined before. If so, the *#ifndef* directive will fail, and the preprocessor will skip over the entire file content, and consequently avoiding duplicate names. Otherwise, it defines the *guard macro* and continues preprocessing the rest of the header file until the *#endif* directive. Alternatively, you can also use the non-standard *#pragma once* directive, if support by your

compiler (e.g., MS Visual Studio 2012 supports it – please try it by yourself).
Continuing with the 'Implementation and Testing' phase, let's then present the remaining programming algorithms, as well as some debugging results in Figure 2.28 and Figure 2.29.

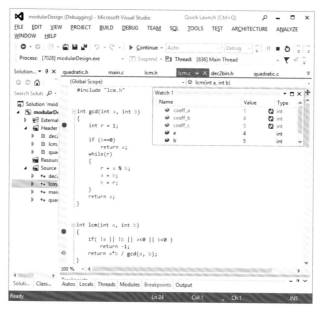

Figure 2.28   Implementation file with the programming algorithms for algorithms presented in Figure 2.20 b) and Figure 2.20 c)

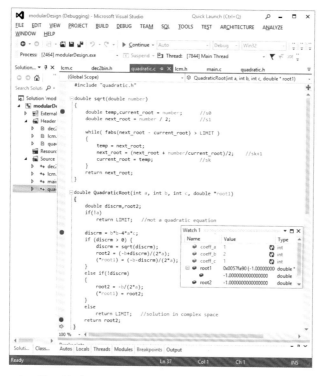

Figure 2.29   Implementation file with the programming algorithms for algorithms presented in Figure 2.9 c) and Figure 2.20 a)

Finally, a programming algorithm for the alternative LCM algorithm on Figure 2.22 is shown in Figure 2.30 to conclude this paragraph.

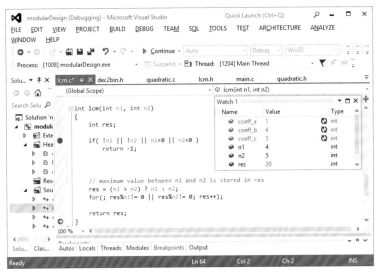

Figure 2.30   Programming algorithm for the alternative LCM algorithm shown in Figure 2.22

2.5   Function Scope and Variable Lifetime

While still in the 'Implementation and Testing' phase, let's try to identify the location of each data variable of the above program, to better understand and follow the debugging process. In doing so, we shall start by quickly describing a generic C program memory layout (a.k.a., C program memory map, C memory model or C program's memory space). This C runtime memory model defines the semantics of computer memory storage for the purpose of the C abstract machine, and it is usually split into the following five memory areas as depicted in Figure 2.31 (excluding the lowest part that is only a single address), all sharing the RAM memory of a computer system and laid out in a predictable manner.

Notice that the data memory segment uses to be further split in several different regions or sections, according to variable size and/or initialization (e.g., the data segment can be seen as a combination of initialized and uninitialized sections). The address 0 belongs to the operating system which uses it to define the NULL pointer (i.e., NULL is a special pointer to memory address 0). Thus, it is not part of any memory segment a user-defined program are allowed to access (e.g., both for read from or write to) and consequently, the reason why you will be reported with a segmentation fault when trying to read from or write to any NULL pointer. This issue will be later addressed on the next chapter, when dealing with dynamic memory allocation, the heap and Arg&Env memory regions.

Let's start by studying and exploring the data segment to locate the global variable, *arrayOf-Coeff*, in the programming algorithm presented in Figure 2.25. However, debugging resource beyond *Watch window* should be on place to provide a wider view into the memory space of the application under analysis. Such resource is denominated *Memory window* and it is able to display everything in the memory space, whether the content is data or code. By default, MS Visual Studio provides you with four memory windows, which you can make visible while debugging the following program, as shown by Figure 2.32.

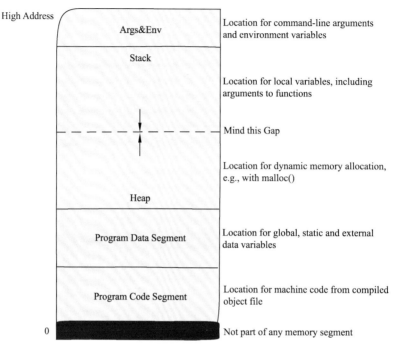

Figure 2.31   A typical virtual memory layout of an executing program. If the program runs out of memory, maybe stack and heap segment will collide and the program will collapse

Figure 2.32   Steps for making a Memory Window view visible from the debug menu, for debugging purpose

As an array, the variable name *arrayOfCoeff* is also an address, and so, type it in the address box of *Memory window* (see Figure 2.33) to view its content (see Figure 2.34). Alternatively, just type the given address at the *Watch window*.

Let's mark in Figure 2.34 the coefficients a, c, and c of the first, second and last quadratic equations, respectively. The bytes in each coefficient are ordered according to the little-endianness of the x86 processor family.

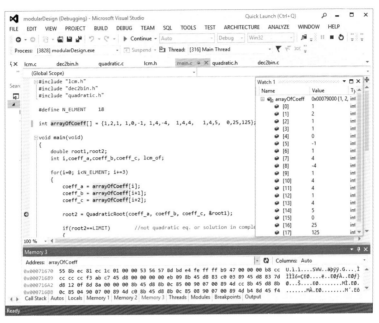

Figure 2.33 Preparing to view the content from the address Of the 'arrayOfCoeff' variable by typing its name in the address box and then press ENTER

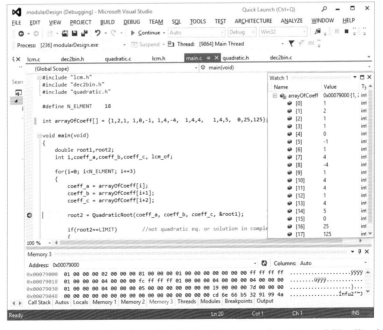

Figure 2.34 Memory Window 3 used to display the content from "arrayOfCoeff" address

By continuing with the debugging process, the execution flow will jump to the *QuadraticRoot()* and the variable *arrayOfCoeff* is still available there (i.e., in the body of *QuadraticRoot()*) as shown at the *Watch* window on Figure 2.35.

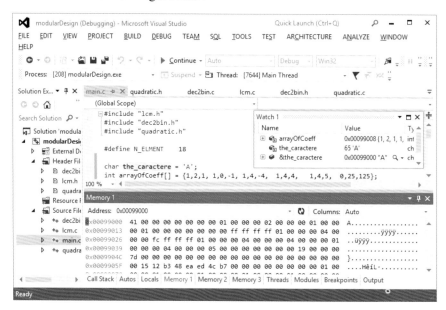

Figure 2.35 Displaying together both global variables, "the_caractere" and "arryOfCoeff" (represented by the first and last elements), just to prove they are both in the data segment

Such visibility will be permanent during the debugging process, even when the execution control is passed to *sqrt()*, *dec2bin()*, and *lcm()*. In this debugged version, *gcd()* is never called as the alternative implementation for *lcm()* was used.

Let's, for example, define another global variable, *the_caractere*, in *main.c* (alternatively declare it in one of the provided .h files, e.g., in *dec2bin.h*) and explore its storage location. To make its location visible in the *Memory window*, type &*the_caractere* in the address box followed by ENTER, and 0x41 = ASCII('A') will show up along with the content of *arrayOfCoeff*. By the proximity (i.e., 8 bytes apart) as shown in Figure 2.35, it is obvious they belong to the same data region, denominated above as data segment. Once again, the variable *the_caractere*, is still available, even when the control is passed to other functions. *What if you try to access one of these variables in another translation unit or source files (e.g., dec2bin.c)*? Try to execute the following statement "*printf("the caractere is %c\n",the_caractere);*" as the last statement of *dec2bin()*. A compiling error will be reported in the *Output window*, as that variable is undeclared in that translation unit. To fix it, just add the following statement "*extern char the_caractere;*" below the include statement in *dec2bin.c* and the linker will solve the unrecognized symbol by the compiler (see Figure 2.36). By preceding the variable, *the_caractere*, with the extern specifier, you are only declaring it (i.e., without defining it), while informing the compiler that such variable is defined somewhere else as global (e.g., in this case in *main.c*).

2 HANDS-ON FUNCTIONS 85

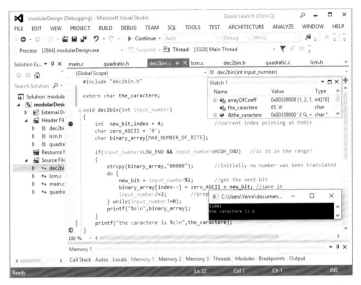

Figure 2.36 Use of the extern specifier to declare a global variable in a different translation unit from the one where it was defined

To conclude these experiments, the following conclusions will be drawn, regarding global variables in C:

1. They are available throughout the program;
2. They are accessible throughout the program if adequately declared;
3. Their locations' storage are all under the same memory regions, the data segment.

Let's now try to identify and explore locations' storage of local variables in C programs. To make such exploration a little bit easier, all local variables including arguments passed to each executed function will be ordered in the *Watch window* in the following way:

1. Following the calling convention, the arguments will be shown from right to left (e.g., for the *lcm()* is first *n2* with higher address and then *n1*);
2. All local variables will be shown by the order they were declared (e.g., for the *QuadraticRoot()* is *discrm* with higher address followed by *root2*);
3. In the *Memory 1 window* you will see the stack content starting at variable with lowest address, i.e., the last pushed on the stack;
4. Beware any time the program is loaded into memory for execution, it'll only preserve the layout, as the operating system' loader can assign a completely different load address. Therefore, the address you will see when repeating this steps will be not the same.

Figure 2.37 to Figure 2.43 illustrate the collected debug data used later to infer some crucial knowledge regarding to storage location, visibility and accessibility of local variables to a C function.

From Figure 2.38 you can also see the stack growing toward lower addresses, as shown by the value of stack pointer (i.e., ESP=16186028=0x00F6FAAC) and also by comparing addresses of local variables while moving execution control from *main()* to *QuadraticRoot()*. The stack pointer changes from 0x00F6FBB0 to 0x00F6FAAC and thus, addresses on the stack of *QuadraticRoot()* local variables became smaller than those of *main()*, as shown by the *Watch window* in Figure 2.37 and Figure 2.38.

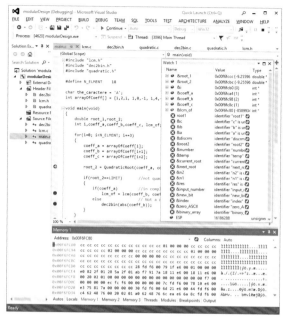

Figure 2.37   While executing main(), only local variables in its body are available with their contents on the stack frame shown in the Memory window

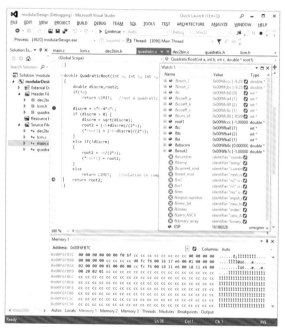

Figure 2.38   When main() transfers control to QuadraticRoot(), only local variables and arguments of the latter become available

When returning from *QuadraticRoot()* back to *main()*, the visibility of *main()* local variables is again activated, while the one for *QuadraticRoot()* local variables is lost. As you can also see from Figure 2.39, the stack pointer is also restored to its previous address before the call site for *QuadraticRoot()*.

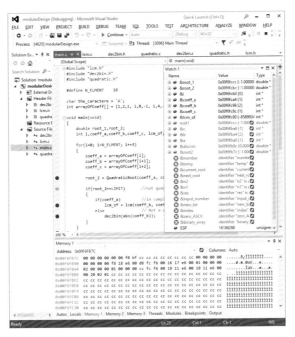

Figure 2.39   Execution control is back to main() and local variables visibility is restored along with the stack pointer

The stack pointer changed from 0x00F6FAAC to 0x00F6F998 and addresses on the stack of *sqrt()* local variables became smaller than those of *QuadraticRoot()*, as shown by the *Watch* window in Figure 2.38 and Figure 2.40. Also from Figure 2.40, you can see that the only available local variables are those belonging to *sqrt()*.

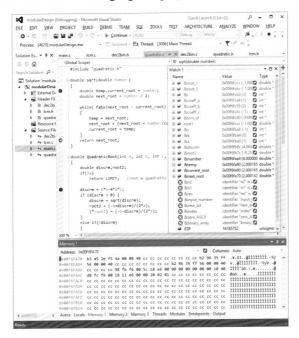

Figure 2.40   Execution control is passed to sqrt() and stack becomes deeper by accummulating local variables from main(), QuadraticRoot() and sqrt()

While passing execution control from *main()* to *lcm()*, the stack pointer changed from 0x00F6FBB0 to 0x00F6FAC4. As it was expected and also presented on Figure 2.41, the only available local variables in the *Watch window* are those related to *lcm()*.

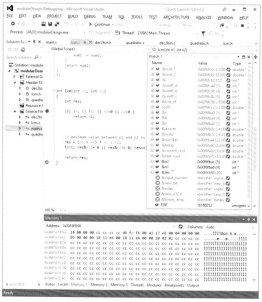

Figure 2.41   While executing the sequence of statement on lcm() body only its local variables will be available

While passing execution control from *main()* to *dec2bin()*, the stack pointer changed from 0x00F6FBB0 to 0x00F6FAA0, as shown in Figure 2.42 with also only local variables related to *dec2bin()*.

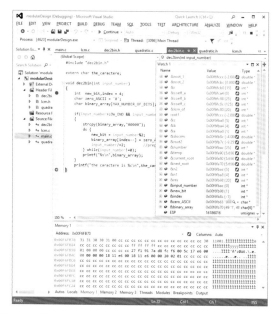

Figure 2.42   Illustration of the stack state related to dec2bin() and also the visibility constrained only to its local variables

Figure 2.43 illustrates the end of program execution with the stack pointer back to its initial address, and also the state of the stack frame related to *main()* execution.

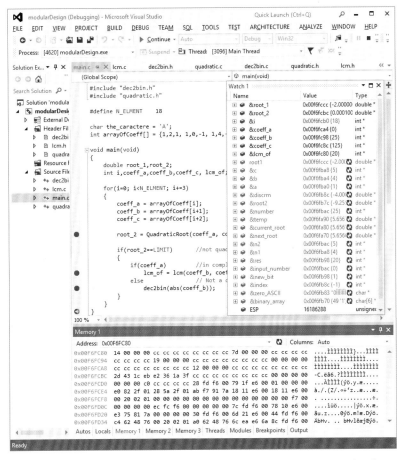

Figure 2.43   Execution control is back again to main() and execution will be terminated with top of stack back to the initial address

To conclude this exploration, the following conclusions were drawn:
1. Local variables and parameters are only available and accessible on the related function body;
2. The stack is a collection of stack frames, each representing a function call. It grows with each call toward lower addresses and it shrinks toward higher addresses as functions return to their caller.

Let's now review the conclusions drawn during the two above debugging explorations at the light of C concepts like scope, lifetime, visibility and namespace, here defined as:
1. The *scope* rules govern whether a piece of code has access to another piece of code or data, and for a particular case of a declared variable, is the area of code where that declaration is in effect (i.e., where the variable is available and can be referenced). For instance, a global variable is available for use throughout the entire program after being declared, while function local variables and parameters are available only in the function body. It

is not allowed to duplicate declaration/definition of two different variables with the same name within the same scope. There are five scopes in C: global, file, function, block, and prototype.

    a. *Global scope* for non-static declared identifiers such as *the_caractere*, *arrayOfCoeff*, *QuadraticRoot*, *lcm*, *gcd*, *sqrt* or *dec2bin* can be accessed or are available anywhere in a program;

    b. *File scope* for static declared identifiers which can be accessed or are available from the declaration site to the end of the source file;

    c. *Block scope* for declared identifiers, including formal parameter, which can be accessed or are available between the opening and closing bracket (*please indicate some from the above program*);

    d. *Prototype scope* for identifiers declared in function prototype, e.g., *positive_number1*, *positive_number2*, *coefficient_a*;

    e. *Function scope* for labels used in a function body as target addresses of goto statements. All variables locally declared always have *block scope and never function's scope*.

2. The *lifetime* of a declared identifier is the time period in which the identifier has valid allocated memory space (i.e., it determines how long a variable storage is guaranteed to still be reserved). For instance, the lifetime of a global variable is the entire duration of the program execution as its storage location is in the data segment, while the lifetime of a local variable is the duration of the function' execution as its storage location is on the stack. C leverages three types of lifetime:

    a. *Static* is also for the duration of the program execution and it requires the use of static qualifier preceding the declaration;

    b. *Automatic* which is until the return of a function or end of a regular block of statements enclosed between { and }, i.e., in this case, the lifetime is limited to the related scope;

    c. *Dynamic* which is managed by the programmer who allocates and deletes pointer variables on the heap at will, using, e.g., *malloc()* and *free()*).

3. *Visibility* is about the "accessibility" of declared variables and is strongly related to the Closest Nested Scope Rule: "*A variable declared has its scope in the current block, and any internally nested block, unless there is a definition with the same name more locally*". It arises when there are declarations' between variables in outer and inner scopes overlap, i.e., variable with the same name in nested scopes. It creates a *scope hole* of the outer variable inside the inner block as the innermost variable takes precedence over the outer variables. For instance, looking at Figure 2.25, the caller (i.e., *main()*) defines a local variable *root1* and has a call site for *QuadraticRoot()*, while in Figure 2.29 the definition of the callee (i.e., *QuadraticRoot()*) has a formal parameter, also named *root1*. In this case, the formal parameter *double *root1* shadowed the *double root1* local to *main()*. Similar scenario happens if you define several variables with the same name in different scopes, e.g.,

one global, another one local to a function, and finally a local to a *while-block* inside the function body.
4. *Namespaces* are named program regions used to limit the scope of variables inside the program and thus, preventing name conflicts in large projects. C language does not support user-defined namespaces but it implicitly partitions program identifiers into the following namespaces, with identifiers in one namespace considered to be different from identifiers in another: (1) tags for *struct*, *union* and *enum*, (2) members of *struct* and *union*, (3) labels, and (4) ordinary identifiers, such as functions' names that are grouped together in a single, flat namespace. There is also a fifth namespace within the preprocessor for macro names and the names of macro formal arguments.

Although, the problem under study was very small and simple for didactic purpose, let's emulate the evolution phase as presented in Figure 1.1, by supposing the following demanded modifications:

1. Solving also the quadratic equation with roots on the complex space;
2. Improving *lcm()* to not repeat full execution if the parameters are the same as the last time it was called. For instance, let's repeat the coefficients (1, 4, 5) twice one after another in the *arrayOfCoeff*.

Based on the new knowledge about scope, you can slightly revise the previous program by removing the prototypes of *gcd()* and *sqrt()* from *lcm.h* and *quadratic.h*, respectively, as both are only used in the source file where they are defined (e.g., define them now as static functions). Let's also keep using the previous implementation of square root only for the roots in real space and *sqrt()* from *math.h* for roots in complex space. In *solveQuadraticEq()*, the local static *sqrt()* will be used while the standard *sqrt()* will be called in *main()* body only for demonstrating the usage of functions with the same name, all of them static except one. Let's also force a convention naming to always assign meaningful name to functions, starting with a verb to indicate action (e.g., *solveQuadraticEq instead of QuadraticRoot*).

Which is the sqrt() called at the call sites in solveQuadraticEq() and why? According the scope rule, it should be the static *sqrt()* as proven by Figure 2.44. From the call site in *main()*, the standard *sqrt()* was called because according to Figure 2.45 the control did not pass through the static one. It was expected because *main()* only see the standard *sqrt()* through the *math.h*, as included in *quadratic.h*.

Can you use debugging to prove the block scope and lifetime of static local variables disc, real and imag? With the switching from available to unavailable (see Figure 2.46 to Figure 2.49) when control is given to *findLCM()* and then returned back to *main()*, it becomes obvious the scope of static local variables, *last_a*, *last_b* and *last_result*, are the body of *findLCM()*. It is also visible from these figures that their lifetime is for the duration of the program execution, as they always keep their values from one call to another.

Figure 2.44   Calling of static sqrt() from the call site in solveQuadraticEq(), according to C scope rule

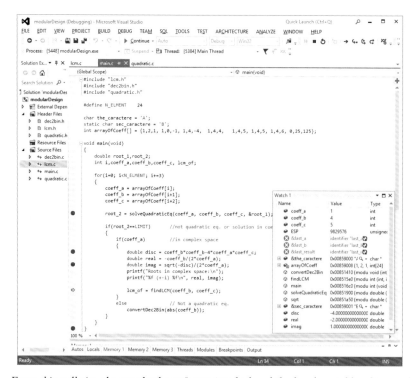

Figure 2.45   From this call site, the standard sqrt() was invoked and the local variables disc, real and imag, inside the if-block became available

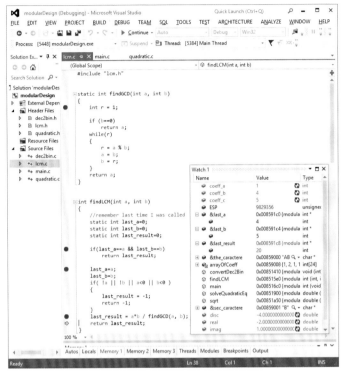

Figure 2.46 The local static variables, last_a, last_b and last_result became available for the first time and values were calculated and then saved for next calls of findLCM(), while disc, real and imag became unavailable again

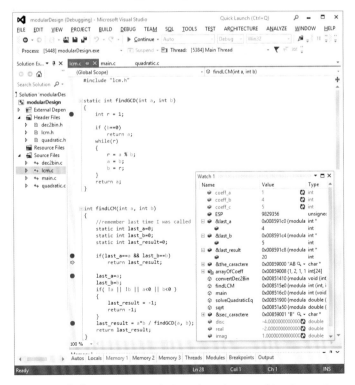

Figure 2.47 findLCM() was called once again and since there is a matching in previous and currently passed coefficients, the last memorized result will be directly return to main()

Figure 2.48   When findLCM() returns control back to main(), the static local variables became again unavailable

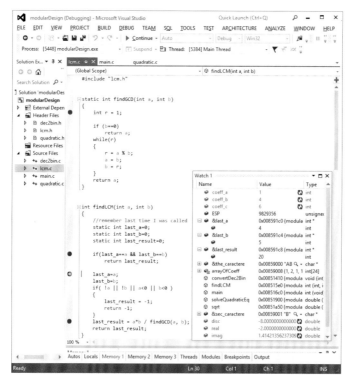

Figure 2.49   findLCM() was called once again and since there is no matching in previous and currently passed coefficients, values were calculated and then saved for next calls of findLCM()

2 HANDS-ON FUNCTIONS 95

To conclude this exploration, the address of all functions were identified by typing the name of each function (i.e., remember that the name of a function is an address) on the address box of the *Memory window* followed by ENTER (e.g., see Figure 2.50). Table 2.5 presents a summary of addresses collected to clearly identify the three regions of RAM memory which correspond to stack, data and code segments. From the grouped addresses it is obvious all stack frames are on the same RAM area, which also happens with function bodies as well as with static and global variables. From Table 2.5, memory segments are laid out from higher to lower address in the following order, stack, data and code, as presented in Figure 2.31.

Figure 2.50 The first byte of code, 0x55, is typically part of the C function prologue code representing the assembly instruction PUSH EBP to save the frame pointer

Table 2.5 Identified addresses used to locate each memory segment

Segment	Symbol	address
Stack for functions	main()	9829576 = 0x0095FCC8
	solveQuadraticEq()	9829316 = 0x0095FBC4
	findLCM()	9829356 = 0x0095FBEC
	convertDec2Bin()	9829304 = 0x0095FBB8
	static sqrt()	9829040 = 0x0095FAB0
	findGCD()	9829124 = 0x0095FB04
Data	last_result	0x008591C8
	last_b	0x008591C4
	last_a	0x008591C0
	sec_caractere	0x00859001
	the_caractere	0x00859000
	arrayOfCoeff	0x00859008

Continued

Segment	Symbol	address
Code	convertDec2Bin()	0x00851410
	findLCM()	0x008515E0
	main()	0x008516C0
	solveQuadraticEq()	0x00851900
	static sqrt()	0x00851A50
	findGCD()	0x00851570

2.6 Recursive Function

Recursion is a particular kind of mathematical reduction (i.e., the rewriting of an expression into a simpler form) which occurs when an entity like an expression, a function or an algorithm is defined in terms of "simpler or smaller versions" of itself. As examples, the factorial of a non-negative integer n and the Fibonacci sequence can be rewritten as "$n! = n*(n-1)!$" and "$F(i) = F(i-1) + F(i-2)$", respectively. It is commonly used as a problem-solving strategy which can be briefly defined through the following steps (1) finding given instances of the problem that are small or simple enough to be directly solved, (2) reducing the problem to one or several simpler instances of the same problem and, (3) figuring out how a solution based on identified smaller instances can be drafted in an evolutionary manner to solve the problem as a whole, while guaranteeing the problem solution finiteness (i.e., avoiding circularity). Putting it differently, one just need to simplify the original problem by finding smaller versions of it, which are solved in similar ways or solve it directly, in case the simplification is unnecessary. Thus, *what are the main components of a recursion as problem-solving strategy?* A recursive definition consists of the following components or elements which must be identified in any problem intended to be recursively solved:

1. *Base cases* are directly and easily solvable instances of the problem with no need for recursion, as the answer is already known (e.g., $0!=1$). They break the chain of recursion and so, they are usually used as conditions for terminating the recursion.
2. *Variant* is any expression or part of the problem that changes (e.g., the problem input value n in $n!$) to simplify or make the problem smaller after applying a recursive case, and so, converging the solution to a base case. For instance, the input value n of a factorial successively and strictly decreases towards zero (i.e., it is bounded below by 0), reaching $0!$ which is a base case.
3. *Recursive cases* when applied should simplify the variant to eventually converge to a base case (i.e., they should drive the problem solution towards points in which the answer is known and recursion becomes unnecessary). For the specific case of $n!$, the computation is successively reduced to $(n-1)!, (n-2)!, \ldots$ and finally $0!$.

Summing up, recursion only works when a problem has the above recursive structure with the entity defined in terms of a variant of itself and also because the variant gets simpler and simpler after successive application of recursive cases or steps, until its solution becomes obvious.

Having the basic concept of recursion defined, then the point is: *what is a recursive algorithm or function*? They can be defined as algorithms that mirrors the above recursion definition, while representing a problem solution by reducing the problem to smaller instances of itself and function that calls itself with smaller or simpler actual parameters, respectively. Algorithmically, a recursive algorithm can be represented by the following flowchart algorithms in Figure 2.51 and Figure 2.52.

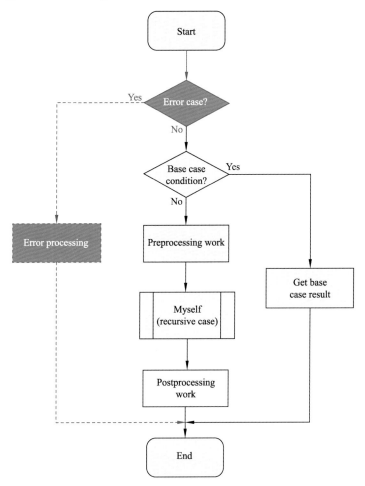

Figure 2.51   General algorithmic structure for recursion, consisting of only one base and one recursive case

The error processing block is optional and it should be removed to avoid its inefficient processing at each recursive case. Figure 2.53 shows how the error case should be processed only once, before applying, e.g., a recursive factorial algorithm.

The analysis and design phases for recursive problems are very specific and simplified if the following methodology is applied. The analysis phase will be mainly based on identifying the above three components of a recursion problem, while the design phase is basically the instantiation, in a refined way, of the algorithms on Figure 2.51 to Figure 2.53. Here is the methodology proposed for analysis and design of recursive algorithms:

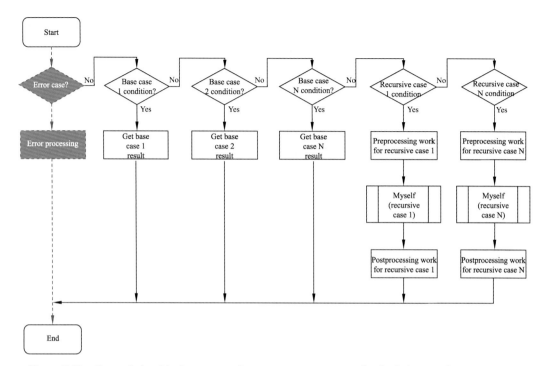

Figure 2.52   General algorithmic structure for recursion, consisting of multiple base and recursive cases

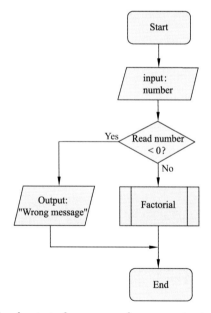

Figure 2.53   Elimination of useless test of error case when processing individually each recursive case

1. Choose for the problem at hand, a variant that consistently bounds to successive smaller ones;
2. Identify and handle error cases to be processed by the caller function;
3. Identify and handle base cases to be used as stopping rules of the callee function;
4. Identify and handle recursive cases by properly and recursively calling them;
5. Tie them all for constructing the final problem solution.

Regarding to the "implementation and testing" phase and more specifically to the writing of a C programming algorithm (i.e., a concrete instance of the algorithms shown in Figure 2.51 or Figure 2.52), one needs to tightly connect one or more base cases to one or more recursive cases together through nested if-else statements. The if-else blocks corresponding to recursive cases must always have recursive calls, while those corresponding to the base cases must always end by returning known values (i.e., no recursive call).

Let's now discuss the analysis and design of a recursive factorial algorithm, mathematically and algorithmically expressed by Equation 2-3 and Figure 2.54, respectively.

$$n! = \begin{cases} 1 & \text{if } n = 0, \\ n*(n-1)! & \text{if } n > 0 \end{cases} \quad (2\text{-}3)$$

Starting with the analysis phase, let's try to identify the variant, as well as the base, error and recursive cases:

Variant: the input n.

Error case: for $n < 0$ an error message is produced.

Base case: for $n = 0$ and $n = 1$, both results are equal to 1 and the recursion ends.

Recursive case: $n > 0$ the result is simplified to $n * fact (n-1)$.

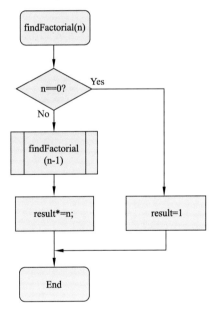

Figure 2.54   Recursive algorithm for n! as an instance of the generic algorithm in Figure 2.51

The algorithm shown in Figure 2.54 is the output of the design phase and it is a refined or fine-grained instance of the generic algorithm presented in Figure 2.51. The "implementation and testing phase" starts by converting the algorithm in Figure 2.54 to its own fine-grained representation as a C programming algorithm in Figure 2.55. This latter figure also presented *main()* as a fine-grained instance of Figure 2.53, as well as the call stack for five calls of the recursive *findFactorial()* plus the calling of *main()*. Since no preprocessing is required, such block was removed from the final algorithm in Figure 2.54. Figure 2.56 shows a graphical representation of the call stack for all recursive calls.

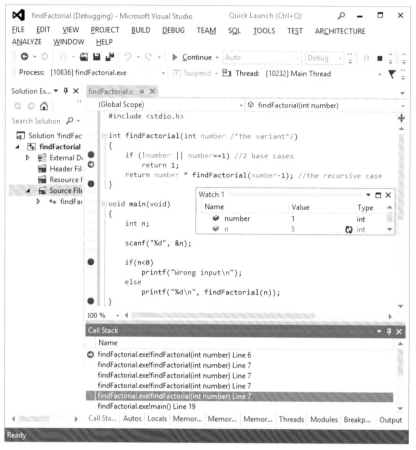

Figure 2.55   Programming algorithms representing recursive findFactorial() and main() with error processing

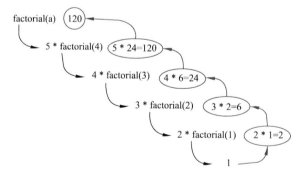

Figure 2.56   Graphical representation of the call stack for 5!

Let's now discuss the design of a recursive *gcd* algorithm, mathematically and algorithmically expressed by Equation 2-4 and Figure 2.57, respectively.

$$\gcd(a,b) = \begin{cases} b & \text{if } a = 0, \\ \gcd(b\%a, a) & \text{otherwise} \end{cases} \qquad (2\text{-}4)$$

By following exactly the same steps as in the factorial problem, let's start with analysis phase: Variant: possible choices are *a, b, b%a* or *a%b* and the point is: which one effectively and consistently reduce the problem in recursive cases? The last two for sure and you should check the other ones.

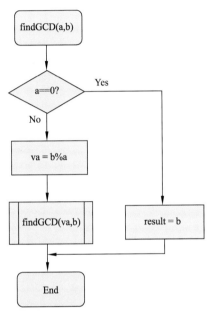

Figure 2.57   Recursive algorithm for gcd(a,b) as an instance of the generic algorithm in Figure 2.51

Error case: for *a* < 0 or *b*<0 an error message is produced.
Base case: for *gcd(0, b)* or *gcd(a,0)*, the result is equal to *b* or *a* and the recursion ends.
Recursive case: $a \neq 0$ the result is simplified to *gcd(b%a, a)*.
The algorithm shown in Figure 2.57 is a refined or fine-grained instance of the generic algorithm presented in Figure 2.51 with no post-processing. Figure 2.58 represents the C programming algorithm for the algorithm shown in Figure 2.57.
To conclude this paragraph, it is worth mentioning that apart from algorithm elegance and efficiency, recursion and iteration technically differ as the stop condition in the former is verified during the program execution (e.g., just take a look at the call stack in Figure 2.55) and not at its end as in iteration. While a base case is not reached, the entire portion of program is called again, as a function of itself, particularly as function of the unfinished original portion of code.
As a final exercise, imagine it is demanded for an update to the previous program discussed at the precedent section 5, to end the *arrayOfCoeff* with a *(0,0,0)* and also to guarantee at the very beginning the array has an exact number of possible quadratic equations.

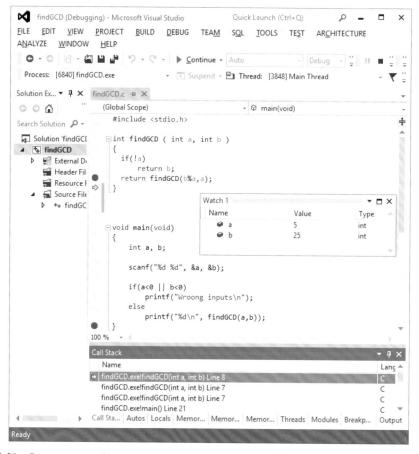

Figure 2.58   Programming algorithms representing recursive findGCD() and main() for error processing

This chapter's main focus was on C function main aspects and programming process as an extended problem-solving process and it will end now with the following recommendations:

Recommendation 10: To improve functional skills toward problem-solving, a lot of practice will be recommended to develop pattern recognition which can be later applied in solving even more complex problems.

Recommendation 11: For large programs use always longer and more self-explanatory parameters' names in a function prototype, while keeping them short and the code concise in function definition.

Recommendation 12: The choice of how to pass arguments to function depends on a case-by-case analysis, while tackling different parts of the program under implementation, and is mostly predetermined by the parameter type or its usage. Otherwise, it is highly recommended to pass arguments by-value to better leverage compartmentalization of data, if arguments are not modified and there is also no performance issue in copying them.

Recommendation 13: Always use header guards construct when your code must be portable to different compilers, as *#pragma once* is a non-standard preprocessor directive, although it offers some advantages, such as less code, no name clashes and compilation speedup.

Recommendation 14: Always assign meaningful names to functions starting with verbs to express actions. It improves code readability and understanding.

Recommendation 15: Although recursion leverages compact and elegant programs compared to iteration-based ones, you should carefully implement them, as they used to spectacularly fail at run-time, mainly due to: (1) missing of base cases, (2) no guarantee of convergence when the chosen variant is not strictly bound to smaller instance of the same problem, (3) excessive memory requirement, mainly concerning with the stack segment, and (4) excessive re-computation.

References

[4] MILLER BRAD, RANUM DAVID. Problem Solving with Algorithms and Data Structures using Phyton. 2th ed. Portland: Franklin, Beedle & Associates, 2011.

[5] AZIZ ADNAN, LEE TSUNG-HSIEN, PRAKASH AMIT. Elements of Programming In terviews in Java. 2th ed. Luxembourg: CreateSpace Independent Publishing Platform PLatform, 2015.

[6] KELLEY AL, POHL IRA, C by Dissection: The Essentials of Programming. 4th ed. Redwood City: Pearson.2000.

[7] HANSEN DEXTER.Flowcharting Help Page (Tutorial). http://www.flowhelp.com/flowchart/.

[8] CHAUDHURI ANIL. The Art of Programming Through Flowcharts and Algorithms. Redwood City: Laxmi Publications, 2005.

Several other information sources were also used, mainly from the internet, as well as those previously referenced in chapter one.

3 HANDS-ON-POINTERS: THE BASICS

Learning objectives

1. Generically understanding pointers and its usage.
2. Understanding and avoiding dangers in returning local pointers.
3. Understanding and practicing how to pass and return structures, strings and arrays to and from functions.
4. Practicing the pointer' usage by examples.

Theoretical contents

1. Pointer and memory.
2. Pointer expressions and arithmetic.
3. Pointers and arrays.
4. Pointers and structures.

Strategies and activities

1. Always exercising each presented topics by first exemplifying along each theoretical introduction and then presenting right away similar problem to be completely and individually solved by students.
    a. Always offers bonus to the first students finishing the exercise as well as to those presenting best solutions.
    b. Bonus should be also offered during all classes starting from the first one.
    c. Choose some of the students with best solutions for briefly presenting their solutions and also helping other students altogether with the instructors.
    d. Give bonus to students revealing certain level of intellectual curiosity on a sequence of two classes, at least.
2. Running a simple quiz about all pointer' knowledge applied and covered in this chapter.

Pointer is an important low-level feature of the C language, mainly because it leverages capabilities to more directly access computer's memory, specifically through addresses at which data are stored within the main memory. Such bare-metal manipulation that C provides for pointer arithmetic, has been differentiated it from other middle-level computer languages like Java or C#. From the programmer perspective, a pointer can be seen as a special kind of variable holding the memory address or location of a program data (e.g., a regular variable) or code (e.g., a function) object. Its power comes from its split personality (i.e., from one side it is a variable storing an address while to the other side it reveals the value at that address), as well as its ability to change values at runtime and enter into arithmetic operations. Due to its low-level nature (i.e., closer to the hardware), pointers are used for a wide variety of purposes, such as (1) providing fast means of referencing array of elements, (2) allowing functions to modify their actual parameter when passed-by reference, and (3) supporting dynamic data structures and memory management. Faster, efficient and flexible code can be generated into machine code by the compiler, eliminating performance overheads incurred by other operators (e.g., array indexing), while leveraging program objects[①] *that change at runtime. However, such advantages, mainly the flexibility, came at a higher programmer responsibility in avoiding new and uglier types of pointer bugs which can crash a program in random ways, making them more difficult to debug.*

3.1 Pointer and Memory

Since pointers contain addresses in memory, understanding them demands for a firm grasp of memory and the functional organization of a C program when loaded into memory, as previously discussed and shown in Figure 2.31. However, not all computer systems use such a pure segmented-memory allocation technique. For instance, modern computer systems (e.g., your desktop or laptop) mix and extend such segmented-technique with a paged memory allocation strategy, in which program code and data are not completely loaded at once, but on demand (i.e., as needed) and in pages (i.e., a fixed-size block of memory) during execution. Previously in Chapter 2, variables and pointers were qualified by 5 generic attributes (i.e., name, type, size, content, location), which are mainly differentiated by their contents and locations. From the programmer point of view, nearly all program objects representing code or data can be expressed by such a 5-tuple, although internally and at runtime all them can be seen as a 2-tuple given by their address/location and content. That is, the name is directly mapped to the address while for size and type, code are adequately generated by the compiler according to the runtime environments of programming language and operating system (OS). Taking a logical view of memory as byte addressable and sequentially arranged (i.e., memory as a monolithic linear array), Figure 3.1 portrays the stack segment laid out as a grid diagram, with cells' values represented as hexadecimal values, while Figure 3.2 depicts how pointers in

---

[①] A program object is a named region of storage which can be referenced by an *lvalue* [3].

memory link to cells' values they point at. Variable contents are stored in memory cells inside the computer's memory using their binary representation and if a variable is larger than one byte, then the address is the location of the least significant byte of the variable, which is the smallest address according to the little endianness of x86 processor family. Reading the *Memory 2* window from left to right and up to down, you will see that (1) variable "*c*" is located at address 0x008FFC8B and it occupies only one memory cell with value 0x54, (2) variable "*x*" is located at address 0x008FFCA0 and it occupies 4 memory cells with value 0x000b, (3) pointer variable "*px*" is located at address 0x008FFCAC and it occupies 4 memory cells with value 0x008FFCA0 which is the address of the variable "*x*", and (4) pointer variable "*pc*" is located at address 0x008FFC94 and it occupies 4 memory cells with value 0x008FFC8B which is the address of the variable "*c*".

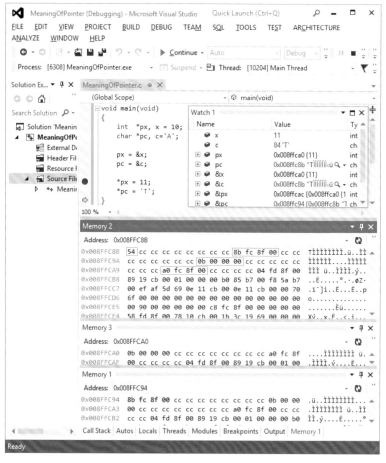

Figure 3.1   Memory, in this case the stack segment, represented as an array of bytes with address as index of a given byte within the array

The point is: why these specific addresses? They are fixed through a collaborative effort among the compiler, the linker and the operating system loading service, taking into account (1) the stack grows from higher to lower addresses (e.g., in this case from 0x008FFCAC to 0x008FFC8B) and (2) the sequential order of variable declarations (e.g., in this case *px*, *x*, *pc* and *c*), and (3) the available RAM memory space at the loading instant. Furthermore, you can see from *Memory 2* window in Figure 3.1 that addresses assigned to *c*, *pc*, *x* and *px* (i.e., in the reverse declaration order), are 0x008FFC8B, 0x008FFC8B+8+1, 0x008FFC94+8+4, 0x008FFCA0+8+4, respectively. The 8s are the unused and padded-like memory cell between each assigned address, which is marked with *0xcc*, indicating the setup/reset value of stack memory cells. *What about the adding with 4 and 1?* Firstly, the size of a pointer depends on the architecture of the CPU and the implementation of the C compiler. Since the MS Visual Studio project was created as a Win32 console application (i.e., it is a 32-bit application), consequently all pointers will have a 32-bit or 4-byte size. Secondly, as you learned in Chapter 1 while playing with pointers and the *sizeof* operator, the size and range of C built-in types will also depend on CPU and compiler architectures, except for variable of type *char* that is always one byte. For variable of type *int* is given by the machine word which, in this case, is 32-bit because win32 console applications are executed under an x86 emulator that allows 32-bit applications (i.e., a virtual 32-bit CPU) to run seamlessly on 64-bit Windows. CPU-dependent is also the alignment of C primitive data type. Data alignment is defined as the memory boundary on which all instances of a type (i.e. variable) should occur and is also given, with few exception [1] , by the *sizeof* operator. Therefore, on a typical x86 32-bit machine, variables of type *char*, *short int*, *int* and *double* should be correctly aligned to memory access granularities of 1, 2, 4 and 8 bytes, respectively. To put it simply, their assigned addresses must be evenly divided by their related access granularities). Furthermore, the pointer alignment is given by the machine word size.

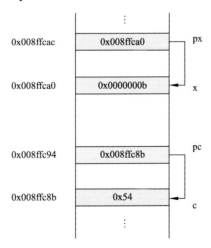

Figure 3.2 The meaning of a variable' content as value (e.g., c=0x54 and x=0x0b) and pointer' content as address (e.g., px=0x008ffca0 and pc=0x008ffc8b) pointing to variable content

One of the most known code object, the function, can also be expressed as a 5-tuple given by its return type, name, content, size and location. Running once again the application and from *Memory 2* window in Figure 3.3, you can easily extract its address, content

[1] On SPARC processors the memory access granularity for *double* type can be 4-byte, instead of the expected 8-byte boundary.

and size. Typing *main* in the address box and then press ENTER, you will see that the name of the *main()* is mapped to the address 0x011C13C0, its content is the set of read-only binary values as red-marked on the figure, and the size is equal to 0x011C144A – 0x011C13C0 + 1= 139 bytes.

Figure 3.4 presents you with another option to see and deeply analyze a program code object (e.g., the *main()*) in terms of its address, content and size, by launching the *Disassembly* window from the *Debug* menu. Additionally, it presents you with assembly instructions generated by the compiler for each source code statement. We shall go back later to this *Disassembly* window to analyze functions step-by-step, while exploring the difference between return from *main()* compared to return from regular functions, as well as different and possible prototypes for the *main()*. It is also visible from Figure 3.4 that the program was once again loaded to another different memory space.

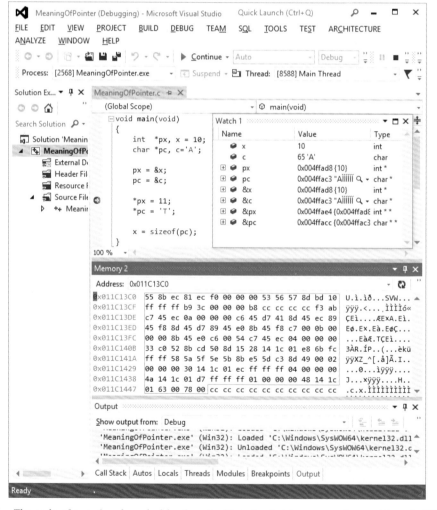

Figure 3.3   The code of main() red-marked by the set of byte loaded to memory from address 0x011C13C0 to 0x011C144B

Let's then conclude this paragraph, calling your attention for the following very important observed facts:

1. A pointer is a kind of reference storing the address of another location in memory, which contains a program object;
2. The size of a pointer variable is always CPU-, compiler- and OS-dependent;
3. Shown addresses during execution can be different at different load times, even on the same computer, as programs are not always load at the same memory space;
4. Pointers and addresses are always treated as unsigned quantities and so, you will see later that only unsigned operations will be allowed to be performed on them;
5. Pointers must be initialized before usage, e.g., as shown in Figure 3.3 by statements "*px = &x;*" and "*pc = &c;*".

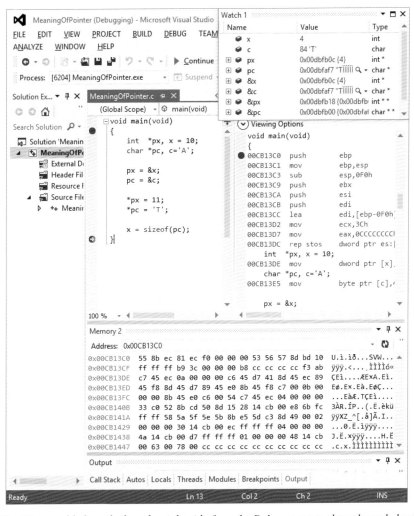

Figure 3.4 The Disassembly launched on the right-side from the Debug menu to show the code loaded to memory from address 0x00CB13C0 to 0X00CB144B, as well as compiler-generated assembly code

3.2 Pointer Expressions and Arithmetic

As said above, pointers are kind of variables used for constructing references, making them fundamental for creating dynamically several and different data structures like recursive data structures, sharing data between different code objects, and offering more flexible flow control when used to point to functions. They are declared exactly as any regular variable except that the name of a pointer variable is prefixed with the address operator (i.e., the asterisk). Simply put, the syntax for a pointer declaration is given by:

base_type *var_name;

It means the *base_type* determines the content at the assigned address and also that the declared pointer must be only used to point to variables of *base_type*. For instance, a "*pointer to double*" or a "*pointer to char*" must point only to variables of type double and char, respectively. *Can you guess what will happen by adding the following statement "pc = &x;" at the body of previous main()?* The compiler will warn you about a type incompatibility as shown by the red mark under the assignment symbol on Figure 3.5. If you try to compile the code, you will see the message "*warning C4133: '=': incompatible types - from 'int *' to 'char *'*". The pointer size is unique but the size of the content of the variables they point at are 1-byte for *pc* and 4-bytes for *&x*. In this case, both warnings do not crash the application but looking at the *Watch window*, the second statement clearly results in meaningless negative value (i.e., a bug) due to the extra 3-bytes adjacent to the address of *pc*. Maybe in another scenario, you might overwrite important data by ignoring such kind of warning.

Figure 3.5 A warning is marked on both assignment statements due incompatible types between pointer variables and variables they point at

Before embarking deeply into the pointer arithmetic, *can you describe the real difference between a pointer and an address?* A pointer can be seen as an address that can be changed at runtime, but addresses are not pointers because they cannot be dynamically changed (e.g., an address mapped by compiler, linker and operating system to any variable name is unique and immutable, like a kind of constant pointer). That is, pointers can be described as addresses that have names and can be stored and later altered, while the addresses assigned to variables has no names or places in memory and so, they are unique and cannot be changed at runtime.

So far, two standard unary operators have been applied on value and address of a pointer variable, the dereferencing operator (*) and its inverse "address-of" operator (&), respectively. Usually, the data or code object referenced by a pointer is called a "pointee" and the dereferencing operator follows the pointer reference to get the value of its "pointee". The dereferencing operator is also known as "value of" operator and is used to both retrieve and store data from and into memory. A flawless usage of pointers starts by generically coding pointer initialization at both pointer- and pointee-levels before the first dereference, i.e., first guarantee legal allocations of pointer and pointee and only then assign the pointer to point at the pointee. In Figure 3.6, the pointer (e.g., *px*), the pointee (e.g., *x*) are statically allocated during their declarations by the compiler and finally linked through the statement "*px = &x;*". Only after the allocation, the first dereference is triggered. Later in the next Chapter 4, you will learn how to perform them all dynamically. For example, the pointer split personality (i.e., looking at a pointer variable differently from the pointer- and pointee-sides) will help you better understanding the sharing promoted during the passing-by reference between a passed argument at caller-side and the correspondent formal parameter at the callee-side, as the two pointers at both sides are both referring or referencing a single and common pointee.

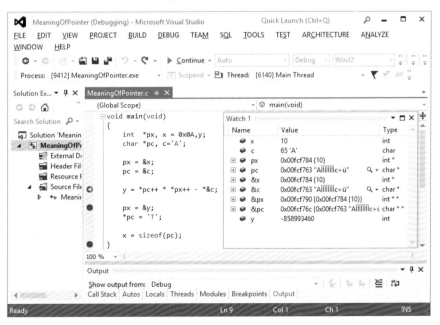

Figure 3.6   Variables and pointers states when the first breakpoint is reached.

Figure 3.6 and Figure 3.7 show that pointers and addresses may be freely used in expressions but under some constraints, as only unsigned operations are performed on them. However, increment and decrement as well as addition and subtraction of integer constant from pointer variable is allowed, but they are recommended only when the original and the displaced addresses refer to positions in the same data object. Otherwise, you should know well how different data objects are laid out in memory, possibly ending up with portability issues. Looking closely to both *Watch* windows, it becomes visible the change of *px* from 0x00fcf784 to 0x00fcf788 and *pc* from 0x00fcf763 to 0x00fcf764. *How do you explain that?* Or *putting it differently, what happens when you increment a pointer?* As has been observed so far, what really differentiates pointers from the other program data objects is the way the pointer is interpreted, and such different interpretation also make pointer arithmetic particularly very different from the regular arithmetic on regular variables. For instance, the value by which the address of the pointer variable is added or subtracted, known as scale factor, is type-dependent (i.e., it scales according to the size of the pointee, ending up with the compiler multiplying that added or subtracted value by the pointee size). Therefore, incrementing one to *px* pointed it to the next integer data object (i.e., it does not necessarily return the address of the next byte after *px*, but the address of the next integer object after *px*), while incrementing *pc* pointed it to the next byte or character. Recall win32 console applications are executed under a virtual 32-bit CPU, i.e., the sizeof *int* type is 4-byte.

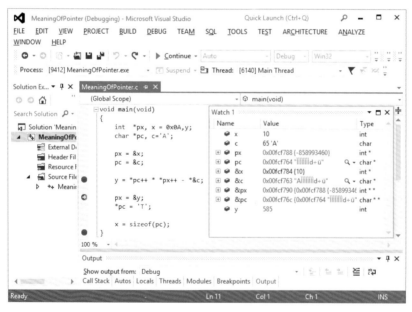

*Figure 3.7 Variables and pointers states at the second breakpoint: the value of y = 65 * 10 – 65 = 585*

Can you figure out the stored addresses on pc and px after executing, e.g., the following two statements "px+=6; pc-=6;"?
Is it allowed to subtract one pointer variable from another pointer variable?
Is it allowed to perform addition, multiplication and division operations on two pointer variables?
Please try them all by yourself, adding new statements to the program and then debugging it.

Looking at Figure 3.7, the asterisk was used in the statement "*y = *pc++ * *px++ − *&c;*" simultaneously as dereferencing and multiplier operators. *How is that possible?* The asterisk is a context sensitive symbol as its meaning depends on where it appears on the statement. Therefore, based on its location in the statement, the compiler will accordingly interpret it as dereferencing operator, as a multiplier operator, or as address operator on type declaration. Furthermore, the dereferencing operator (*) and its inverse *"address of"* operator (&) have higher precedence than all other arithmetic operators except the unary minus, with which they share the same precedence.

3.3 Pointer and Arrays

Since you played long before with array and have also been introduced in previous paragraphs to pointer, pointer arithmetic and expressions, let's then explore the relationship between array and pointer by solving the following problem: "*Compare a binary number, represented separately by its several nibbles, to the largest number found in a sequence of numbers*".

Starting with the analysis phase, let's constrain the solution space to 1-byte positive binary numbers (i.e., a number consisting of only *the most and least significant nibbles*) but later you will be challenged to extend the solution to signed multiple-bytes binary numbers and so, deciding for a strategy to define a known nibbles' sequence, as well as deciding for the bit-signal (e.g., the most significant bit of the most significant byte). Since a nibble is defined as a half of a byte (i.e., a 4-bit number), let's then define two integer arrays of 4-integers each (e.g., *MSn[4]* and *LSn[4]*) to represent the most and least significant nibbles of the binary number. Before the comparison with the largest number of a sequence of number, the binary number has to be converted to its decimal representation. The easiest way to realized such conversion is to accordingly assign to each indexed bit from, the least significant bit *LSn[0]* to the most significant bit *MSn[3]*, its weight as an indexed power of two, and then calculate the accumulative sum of each bit value multiplied by its weight, as given by equation (3-1).

$$Dec(MSn,LSn)= MSn[3]*2^7+ MSn[2]*2^6 + MSn[1]*2^5 + MSn[0]*2^4$$
$$LSn[3]*2^3+ LSn[2]*2^2 + LSn[1]*2^1 + LSn[0]*2^0 \qquad (3\text{-}1)$$

Let's also assume that the sequence of numbers will be stored in an array of *n* known numbers starting from position *i* to position *l* (e.g., *S[i,l]*), which will be fully scanned to find the biggest stored number. Although, a repetition-based problem-solving approach sounds great for carrying out such scanning, let's analyze it following a recursion-based approach (i.e., identifying the variant as well as error, base and recursive cases).

1. Variant: since the idea is successively reducing the inputted sequence to shorter ones (i.e., smaller subsequences), variants are the inputted sequence and its original size.
2. Error case: the sequence should have at least one number.
3. Base case: subsequence of only one number as the largest number is directly calculated as exactly that number.
4. Recursive case: the sequence is successively reduced in one number toward the last

number, and then the maximum is chosen between the currently removed element and the new formed subsequence, as given by Equation (3-2).

$$L(S[i,l]) = \begin{cases} S[i] & \text{if } i = l, \\ \max(S[i], L(S[i+1,l])) & \text{otherwise} \end{cases} \quad (3\text{-}2)$$

There is also a constraint related to the binary number as the value of each bit on both nibbles can only be 0 or 1. Furthermore, no comparison should be performed between the converted number with the biggest one, if the latter is bigger than 255. The 1-byte binary number is always smaller than $2^8 = 256$.

Now, moving to the design phase, 2 modules were identified according to their unique functionalities for: (1) converting from binary to decimal and (2) finding the largest element in a sequence of number. Flowchart algorithms from Figure 3.8 to Figures 3.11 represent a possible problem solution for the problem under study. Once again, the recursive flowchart algorithm in Figure 3.8 is an instantiation of the generic algorithm structure for recursion, which consists of only one base and one recursive case as shown in Figure 2.51. It was refined according to the identified variants, base and recursive cases, as well as the post-processing step which choose and memorize the greatest value.

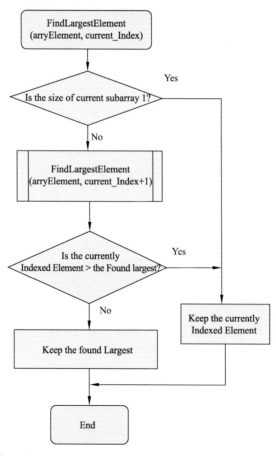

Figure 3.8 A flowchart algorithm expressing recursively the searching for the biggest number in a sequence of numbers

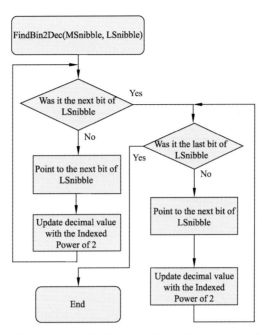

Figure 3.9 A flowchart algorithm expressing the binary to decimal conversion of a 1-byte binary number represented by the most and least significant nibble

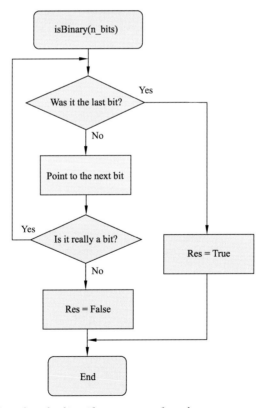

Figure 3.10 A flowchart algorithm checking if a sequence of numbers can represent a binary number (i.e., all them should be only 0 or 1)

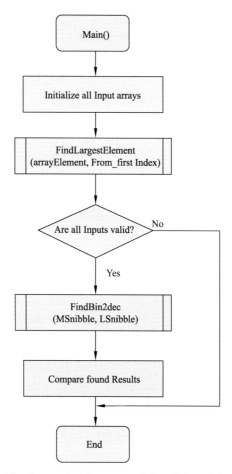

Figure 3.11 A flowchart algorithm integrating the two modules while validating all inputs and comparing the converted number to the largest one

Stepping into the "Implementation and Testing" phase, let's first map analysis phase' entities to chosen program' objects entities as shown in Table 3.1 to make code understanding easier. Shorter symbol names were used during analysis for better representation of equations.

Table 3.1 Mapping of Analysis' phase entities to program object entities

Analysis Phase' Entities	Program Object Entities
MSn	MSnibble
MSn	LSnibble
$Dec(MSn, LSn)$	findBin2Dec
$S[i, j]$	arrayElem[]
$S[i]$	arrayElem[i]
$L(S[i, j])$	findLargestElement

j = NUMBER_OF_ELEMENTS-1 is the right-end index of S[i, j]

Among possible suitable data structures to organize sequences of numbers and bits, array was chosen because it is appropriated to store and processing a list of data items of the same type (e.g., *int* in this case). Before exploring how an array is laid out in memory, let's present the header files with function prototypes and arrays lengths in the Table 3.2.

Table 3.2. Header files for the two identified modules for conversion from binary to decimal and finding the biggest element.

bin2dec.h	largest.h
#ifndef BIN2DEC_H_	#ifndef LARGEST_H_
#define BIN2DEC_H_	#define LARGEST_H_
#define NIBBLE_LENGTH 4	#define NUMBER_OF_ELEMENTS 6
int isBinary(int *const);	int findLargestElement(const int arrayOfNumber[], int currentIndex);
int findBin2Dec(const int nibble_1[NIBBLE_LENGTH], const int nibble_2[NIBBLE_LENGTH]);	#endif LARGEST_H_
#endif BIN2DEC_H_	

Figure 3.12 illustrates the initial execution of the *main()* until the first breakpoint, while showing in *Watch1* and *Memory 1* windows red-marked cells' value of the three declared arrays in memory. As you can see from these windows, for each array declaration the compiler allocated contiguous cells for *int* data objects, along with a constant named pointer referring to the first cell of each array. For example, the *arrayElem[6]* is a named storage region of 6 integer cells laid out from 0x008FF960 to 0x008FF977, with the first cell pointed by a constant pointer to *int* named *arrayElem* (i.e., declaring an array is the same as declaring a constant pointer given by the array name, pointing at the base address of the array).

What do you think will happen by uncommenting the two statements using the named pointer arrayElem in Figure 3.12? Please try it and you will see the compiler claiming for a modifiable *lvalue* on the second statement, exactly because the name of the array is a constant pointer. Thus, you cannot change the memory address it points at. The first commented statement can also be replaced by "*int *p = &arrayElem[0];*" as the constant named array pointer references the base address of the array. Thus, **arrayElem* is the same as *arrayElem[0]*.

Figure 3.13 presents the programming algorithm for the recursive flowchart algorithm in Figure 3.8, as well as the program execution state when the *base case* is reached. Processing the last element of the original array means the current reduced subarray has size equal to 1 as *currentIndex* is equal to 5, which is also the last index of the array, i.e., L(S[5, 5]) = S[5] = 100. After the execution of the base case, successive backtracking through previous recursive calls will be performed until reaching the first call which processed the whole array, starting from the index 0, i.e., L(S[0, 5]) = max(S[0], L(S[1, 5)). From the *base case* execution point, you can successively press F10 for step-by-step execution, while observing the values of *theLargest*, *currentIndex*, and *arrayOfElem[currentIndex]* changing at last to 255, 0 and 8, respectively, as showing in Figure 3.14.

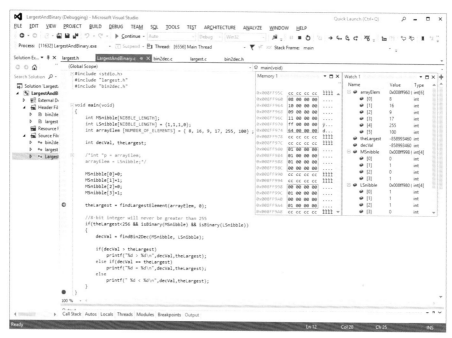

Figure 3.12   Programming algorithm for the flowchart algorithm of Figure 3.11 along with the organization of array elements in contiguous memory cells

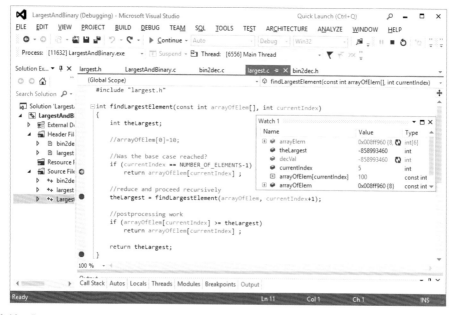

Figure 3.13   Programming algorithm for the recursive flowchart algorithm of Figure 3.8 along with the execution of the base case

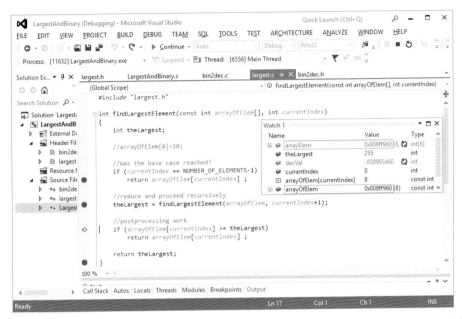

Figure 3.14   Programming algorithm for the recursive flowchart algorithm of Figure 3.8 along with the result from successive backtracking through recursive calls till reaching the first one

Looking at the declaration of findLargestElement() in in Figure 3.14 or on Table 3.2, while rewinding to the parameter passing mechanism previously studied in Chapter 2, could you guess how is arrayElem[] passed to findLargestElement() at the call site in Figure 3.12? According to the relationship between arrays and pointers described above (i.e., array as a contiguous successive objects of the same type pointed by a constant named pointer), the first guess should goes toward pass-by reference. That is right, by default the constant named pointer, *arrayElem*, is passed to a function instead of passing every single value stored in the array. Looking at the *Watch 1* window, the *arrayOfElem* at callee-side shares exactly the same address, 0x008FF960, with the *arrayElem* at the caller-side, as expected due to the default convention call [1] on MS Visual Studio. However, you also learned that pass-by-reference breaks a little the compartmentalization of data leveraged by functions. So, *what if you want the argument arrayElem[] to be read-only?* That is exactly the reason why the *const* qualifier was used to prevent the callee *findLargestElement()* from changing the cells value of *arrayElem[]*. Uncomment the statement "*arrayOfElem[0]=10;*" in Figure 3.14 and you will see the compiler claiming for a modifiable *lvalue.* Then remove the *const* qualifier from the *findLargestElement()* declaration and you will see that everything will be fine.

Alternatively to the use of formal parameters as a sized array as done on the declaration of *findBin2Dec()* and *findLargestElement(),* an array can also be passed using formal parameters as a pointer, as done on the declaration of *isBinary()* shown in Figure 3.15.

---

[1] By default *MS Visual C*/C++ uses the compiler option /Gd for the default, _cdecl, calling convention at translation unit level. Furthermore, one can also declare individually each function for a specific calling convention by properly qualifying its declaration, e.g., "*int _stdcall fun(int, char, char);*". For GNU GCC the _attribute_ qualifier is used, e.g., "*int fun(int, char, char) __attribute_ ((stdcall));*".

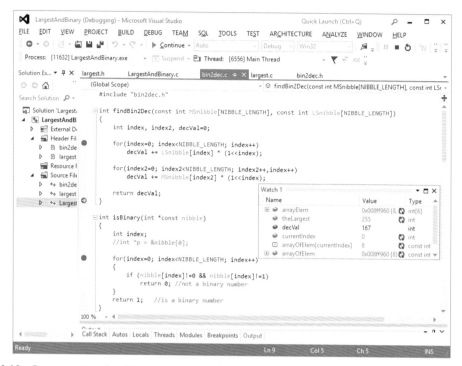

Figure 3.15   Programming algorithms representing the conversion binary to decimal module according to flow-chart algorithms in Figure 3.9 and Figure 3.10

How do you read the pointer declaration as formal parameter of isBinary() in Figure 3.15? To make pointer declaration easier to understand, always read it backwards (i.e., "nibble is a constant pointer to integer data objects"). Reading "*int *const nibble*" backward and progressively, it will be even easier to understand:

1. nibble is a variable argument;
2. nibble is a constant variable argument;
3. nibble is a constant pointer variable argument;
4. nibble is a constant pointer variable argument to integer;

Can you explain why the compiler claims for an initializer in the statement "int * const q;" while it is fine with the statement "const int * q;"? Try to read both statements as above and you will understand why: in the former statement *q* is a constant pointer variable to integer data objects, while in the latter *q* is a pointer variable to constant integer data objects.

Table 3.3 presents and discusses three alternatives programming algorithms for the *isBinary()* according to the flowchart algorithm in Figure 3.10. The first two versions present a third declaration method to pass array as function argument through the use of unsized array as formal parameter. Furthermore, alternative implementations showed that due to the described relation between array and pointer, objects of an array are always indirectly accessed through the use of dereferencing and "address of" operators, e.g., the compiler always transforms *nibble[i]* in **(nibble+i)*. Viewing array as a constant pointer is more primitive and closer to the hardware, while viewing as array is more abstract and sophisticated. Thus, with no optimization concern the choice is up to you.

Table 3.3. Alternative implementations of isBinary() shown the relation between the array indexing operator (i.e., []) and pointer expression

Alternative versions isBinary()	Comments
| ```
int isBinary(int nibble[])
{
  int index, *pNibble = nibble;
  for(index=0; index<NIBBLE_LENGTH; index++)
  {
    if ( *(pNibble+index)!=0 && *(pNibble+index)!=1 )
      return 0;      //not a binary number
  }
  return 1;          //is a binary number
}
``` | 1. Used of formal parameter as an unsized array.<br>2. Compared with version in Figure 3.15, it shows that array indexing is equivalent to pointer arithmetic. So, *nibble[index]* can be converted into a pointer expression as *nibble[index]* is exactly equivalent to *(pNibble+index)* |
| ```
int isBinary(const int nibble[])
{
  int index;
  for(index=0; index<NIBBLE_LENGTH; index++)
  {
    if (nibble[index]!=0 && nibble[index]!=1)
      return 0;      //not a binary number
  }
  return 1;          //is a binary number
}
``` | Used of formal parameters as a constant unsized array |
| ```
int isBinary(int * const nibble)
{
  int *pNibble;
  for(pNibble = &nibble[0]; pNibble<nibble+NIBBLE_LENGTH; pNibble++)
  {
    if ( *pNibble!=0 && *pNibble!=1 )
      return 0;      //not a binary number
  }
  return 1;          //is a binary number
}
``` | 1. Array indexing can be completely replaced with pointer arithmetic, automatically applying the scaling factor, which is 4 in this case, while summing constant to the pointer.<br>2. The *"address of"* operator used to get the base address of the array that is also given by the array name, i.e., *pNibble = &nibble[0]* is exactly the same as *pNibble = nibble* |

The flowchart algorithm shown in Figure 3.16 is an alternative for the one illustrated in Figure 3.9. It merges the two loops which separately process the most and least significant nibbles, as they both iterate over the same index range without depending on each other's specific data. It is a known compiler optimization designated as loop jamming (a.k.a., loop fusion) aiming to reduce loop overhead in order to improve performance and code memory footprint. However, the gain in performance is not always guaranteed in modern cache-based processor architecture, as there are cases in which multiple loops can leverage more data locality within each individual loop, and so, more performance gain opportunities.

Figure 3.17 presents the programming algorithm for the jamming flowchart version of converting a binary to decimal shown in Figure 3.16.
```

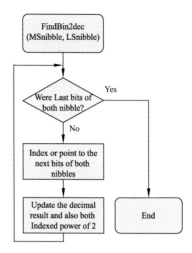

Figure 3.16 Loop jamming version for calculating the decimal value of a binary number represented by its two nibbles

Figure 3.17 Programming algorithms representing the conversion binary to decimal module according to flow-chart algorithms in Figures 3.16 and 3.10, with the latter totally pointer-based

Figure 3.8 to Figure 3.11, Figure 3.16 and Figure 3.18 are too coarse-grained compared to their related programming algorithms, shown in Figure 3.12 to Figure 3.15 and Figure 3.17. *Thus, could you draw their refined flowcharts closer to the final programming algorithms?*

Imagine that after being used for a while, users of the previous program demand for more flexible way to input the binary number, for example, to easily accept bigger numbers than 255. Therefore, the previous problem statement was slightly change to constrain inputted binary numbers as strings. The new revised problem statement will be: "*Compare a binary number, represented as a binary string, to the largest number found in a sequence of numbers"*. The moving to the evolution phase requires short addenda to previous phases, starting with analysis and design phases:

1. During analysis, the concept of string needs to be understood and defined. Therefore, the binary number will be inputted in only one array of ASCII character ended by a special end of string character (i.e., the character '\0').
2. In the design phase *findBin2Dec()* and *isBinary()* need both to be changed to process ASCII characters until reaching the special end of string. Furthermore, each ASCII bit must be converted to its decimal representation. Figure 3.18 presents a new coarse-grained flowchart algorithm for converting from binary to decimal, performing each bit validation during the conversion process (i.e., it merges both functionalities in just one). This new algorithm returns –1 if the inputted string is not binary.

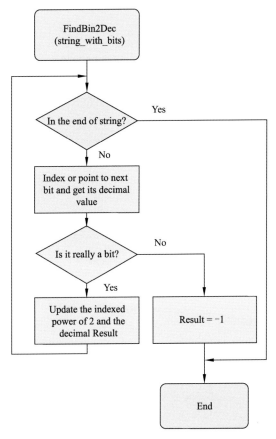

Figure 3.18 A flowchart algorithm expressing the binary to decimal conversion of a binary number represented by a string

Looking at Figure 3.19 and Figure 3.20, you will see that after each iteration, the string pointed by the pointer variable *pBits* will be smaller and smaller, till becoming completely empty (see Figure 3.21). Therefore, it seems a good candidate for a recursion-based algorithm. *Could you redraw the flowchart algorithm shown in Figure 3.18 to an equivalent recursive flowchart algorithm, by identifying first the variants as well as the base and recursive cases?*

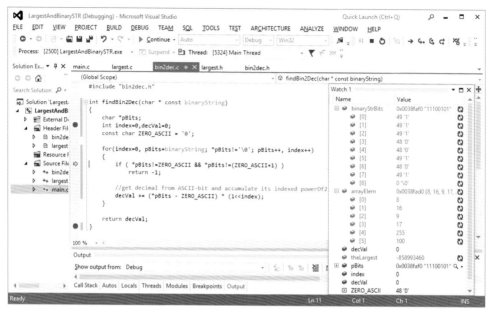

Figure 3.19   Programming algorithm representing the conversion binary to decimal module according to the flowchart algorithm in Figure 3.18 along with the execution state, while validating and converting the first bit of the string-based binary

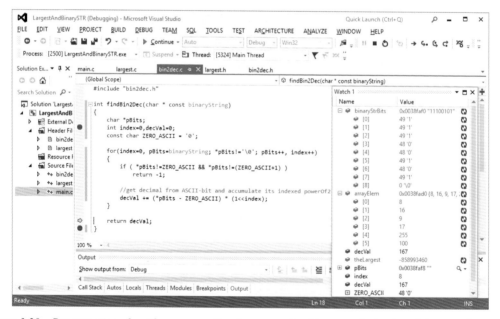

Figure 3.20   Programming algorithm representing the conversion binary to decimal module according to the flowchart algorithm in Figure 3.18 along with the execution state, while validating and converting the last bit of the string-based binary

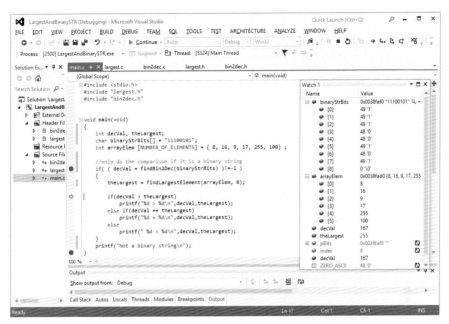

Figure 3.21   The main() was slightly modified to introduce the new algorithm for converting binary string to decimal module according to the algorithms in Figure 3.18 and Figure 3.19

3.4   Pointers and Structures

As described in Chapter 1, a structure data type is one kind of compound user-defined data type used as a form of a block of data. Contrarily to array, it groups together a collection of member or field elements altogether with their probably different data types under one common data pack template. Thus, a structure variable can be seen as the name of a block of one or more variables which may or may not be of the same data type.

Previously also in Chapter 1 you practiced a little with structure declarations and their fields' access through the use of dot operator (.), as shown in Figure 1.41. It was obvious that fields of structure variables are used just like any scalar variables in a program.

Now you will practice another field accessor operator, the "*point-to*" operator (->), also known as arrow operator, to qualify a pointer structure variable name followed by the name of the field you want to access. However, you are still able to dereference the pointer first and then apply the dot operator to appropriately access fields of the structure data object. In doing so, parentheses are needed (see Figure 3.22 and Figure 3.23) because the dot operator has higher precedence than the dereferencing operator (*). The last set of statements shows how to access fields of both pointer structure and structure variables for read and write purposes. Looking also at the last entry on the *Watch 1* window, you will see that the address of a structure variable is the same as the address of its first field (e.g., the storage for *s1* starts on the address of the field named *id1* according to the template or structure type *struct1_t*). However, beware *&s1* and *&(s1.id1)* are completely different pointer types. Contrarily to array declaration, no constant base pointer is assigned to a structure variable declaration by the compiler.

Figure 3.22   Execution till the second breakpoint showing explicitly 2 structure' templates with the same fields, but different sizes of 16 and 20 due to different memory layouts

Figure 3.23   Step-by-step execution from the second breakpoint to the third one, illustrating accessing fields of structure variables using dot and "point-to" operators

How do you explain the differences in size and storage layout of the two structure variables, s1 and s2, knowing they both share exactly identical fields according to their structure type, struct1_t and struct2_t, respectively? Let's start by enumerating some visually observed facts from Figure 3.23 such as:

1. Each structure field is aligned according to its type size which causes a kind of intra-structure alignment, e.g., *id1* (0x00DFFEA4 and 0x00DFFEBC), *id2* (0x00DFFEA8 and 0x00DFFEC4) and *val* (0x00DFFEB0 and 0x00DFFECC) on addresses evenly divided by 4, while *short1* (0x00DFFEAE and 0x00DFFEC8) on address evenly divided by 2 and *char1* (0x00DFFEAC and 0x00DFFEC0) can be aligned on any address;
2. Cells' values of memory addresses 0x00DFFEAD and 0x00DFFEC1 to 0x00DFFEC3 are padded to align *s1.short1* and *s2.id2*, respectively;
3. Cells' values of memory addresses 0x00DFFECA and 0x00DFFECB are padded to align *s2.val;*
4. Enough contiguous memory locations are always allocated for structure variable fields and it is at minimum the sum of the size of individual fields.

Therefore, the differences on size from expected 15-bytes (i.e., 3*4-bytes for 3 *int* fields, plus 1-byte for *char1* and 2-bytes more for the *short1*) to 16- and 20-bytes are due to the padding mechanism to align data types according to their natural alignment (i.e., aligned to their corresponding memory access granularities). The differences in terms of layout are also due to the natural alignment and consequently due to the physical organization of the main memory. Before a short explanation regarding the impact of the physical organization of the main memory, let's answer the following question: *is there any way programmers can reduce the padding?* For a structure with these fields on a 32-bit platform with self-aligned types, padding is unavoidable but it can be minimized if the programmer declares those structure fields in their increasing or decreasing order of size.

Figure 3.24 presents a new structure type, *struct3_t*, with all fields declared in their decreasing order of size with padding at the end of *s3* storage. The occurrence of padding at the end of structure until its stride address can cause a kind of inter-structure alignment, as shown in Figure 3.25. The stride address of a data structure is the first next address of another data object with the same alignment, following the structure storage. From the above observation comes the following question: *is there any kind of structure alignment requirement?* The answer is yes and it is dictated by the compiler which naturally aligns data structures according to their largest field type.

Figure 3.26 illustrates how structure padding can be avoided using *#pragma pack* directive to arrange memory storage for structure fields, ignoring completely the processor's natural alignment rules. *#pragma pack(n)* simply sets the new alignment, and in the above example it is set to memory access granularity of 1-byte (i.e., the same as no alignment). Alternatively, with GCC you can also use _attribute_((packed)) on a structure type to replace #pragma pack(1). However, violating the natural alignment of a processor should happen only under a controlled-way, as it forces the generation of more expensive and slower code, as next you will know why by understanding the physical organization of the main memory in modern computer systems. One of the few possible reasons for using *#pragma pack* is to match your data layout to some kind of bit-level hardware or protocol requirement (e.g., memory-mapped hardware port). Looking at *Watch 1* and *Memory* windows, you will see new sizes and layouts for structure variables *s1* and *s2*. Thus, to avoid storage misalignment, the compiler along with the CPU introduce natural alignment to all scalar and user-defined compound types. For the latter, storages are bound to largest fields type with stricter alignment requirement. For example, try and study for yourself a structure type with arrays and another structures as field types.

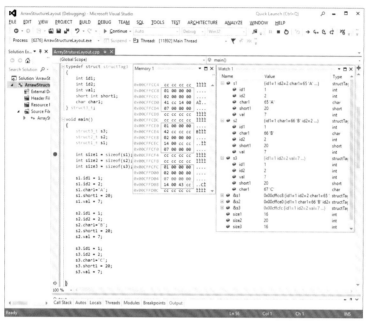

Figure 3.24   Declaring structure fields in their increasing or decreasing order of size to reduce the padding: fields of struct3_t were declared in their decreasing order of size, resulting in just one-byte padding

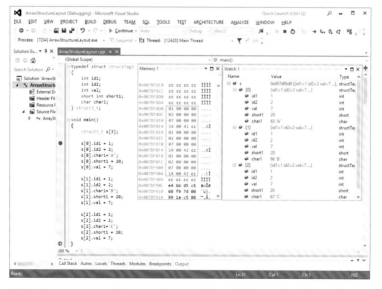

Figure 3.25   Inter-structure alignment due to padding till the stride address of s[1] and s[2]

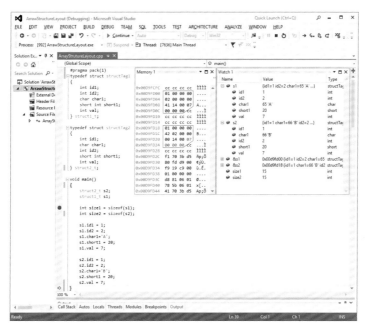

Figure 3.26 Avoidance of structure padding with #pragma pack directive, making the structure size exactly equal to the sum of individual fields' sizes

Now, let's go back to the former question and answer: *how does the physical organization of main memory impact the natural alignment?* So far, to simplify the understanding of storage layout of program data objects, memory has been introduced only from the logical view as a monolithic byte-addressable linear memory (i.e., organized as a singly dimensioned array of bytes). Although easily understandable, such a logical view ignores several and very important issues, e.g., the impact and reason for strict natural alignment of scalar and user-defined compound types by the CPU and compiler, respectively. In spite of many advantages of byte addressability, mapping the memory logical view directly to a single RAM chip, imposes sequential access, e.g., requiring *n* clock cycles to read a word, fetching one byte a time, with *n* equal to the size of the machine word. To alleviate such performance degradation, main memories on modern computer are physically built from a collection of RAM chips with addresses interleaved across them. That is, putting aside issues related to virtual addressing, physically the main memory are organized as byte-addressed contiguous rows of words and accessed a row at a time, as shown in Figure 3.27.

The *Chip Enable* signal will be accordingly generated after fetching and decoding memory access instructions for accessing 1-byte, 2-bytes or a full word. Notice the memory addressing is still sequential, but the read/write of different RAM chips are overlapped and in turn, with the latency of the first bank. For example, accessing an integer variable allocated at address 0x00000100 (i.e., 4), means that bank 0, bank 1, bank 2 and bank 3 will point at 0x00000100 (i.e., 4+0), 0x00000101 (i.e., 4+1), 0x00000110 (i.e., 4+2) and 0x00000110 (i.e., 4+3) addresses, respectively. Figure 3.28 depicts an unaligned memory access from address 7, spanning across two machine word boundaries or rows and requiring two memory read cycle to fetch the data.

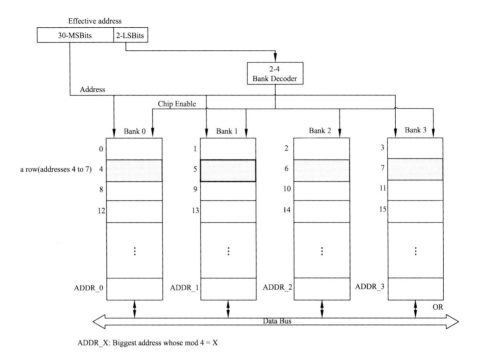

Figure 3.27   4-way low-order memory interleaving with the 2 least significant bits and higher-order bits used to select the RAM Chip banks and the desired data object, respectively

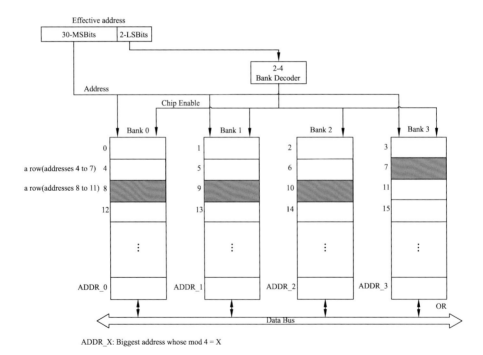

Figure 3.28   Reading from the unaligned address 7, which requires accessing 2 rows and then oring bytes accordingly on the data bus

Understanding the memory physical organization also makes obvious the special case of a 1-byte alignment for char type, making char variables equally expensive from anywhere they live inside a machine word. Furthermore, it is obvious why single in-line memory module (SIMM) or dual in-line memory module (DIMM) are composed by low-order interleaving, which also leverages the principle of locality by reading in parallel adjacent storage locations. Some motherboard chipsets describe a hybrid interleaving with low-order for SIMM/DIMM and high-order for memory banks. High-order interleaving allows memory banks to be accessed independently by different units, e.g., CPU0, CPU1 and Hard Disks, while each of them can use different memory banks. In doing so, higher parallelism will be promoted and consequently higher performance will be achieved.

Now, suppose the users of our previous program demands for new features, for instance, identifying who did ask for conversion and comparison of binary and decimal numbers. The new revised problem statement will be: *"Compare a binary number, represented as a binary string, to the largest number found in a sequence of numbers, as required by a user"*.

In term of analysis phase activities, some way will be devised to register the user responsible for the conversion-comparison request, while in terms of design phase, a specific module will be added to strictly process a request. Figure 3.29 illustrates a coarse-grained flowchart algorithm for request processing. The "implementation and testing" phase starts with an external declarations of structure templates and prototypes in a separate header file, as shown in Figure 3.30, to facilitate consistent usage of tags and functions. This header file is then made part of the two following source files through the *include* directive. A structure type, *structRequest_t*, is proposed to group all variables related to a single request. This phase continues with other activities such as:

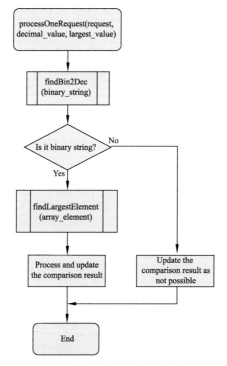

Figure 3.29   A coarse-gained flowchart algorithm for a request processing

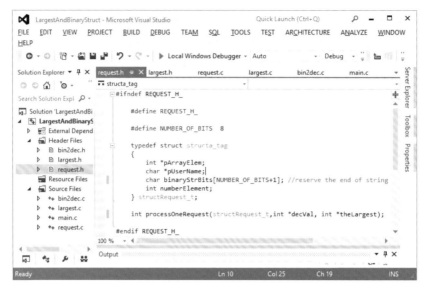

Figure 3.30   External header file supporting the implementation of conversion-comparison request by grouping request-related variables in a user-defined structure type

1. The previous *main()* is refactored as illustrated by Figure 3.31 to call a new function, *processOneRequest()*;

Figure 3.31   Execution of the new refactored main() till the second breakpoint before, calling processRequest()

2. The prototype of *findLargestElement()* is slightly modified to receive another parameter, *nElement*, as illustrated on Figure 3.32;

Figure 3.32   Execution of the programming algorithm, findLargestElement(), after the change of its prototype

3. The flowchart algorithm shown on Figure 3.29 is coded as the programming algorithm depicted on Figure 3.33.

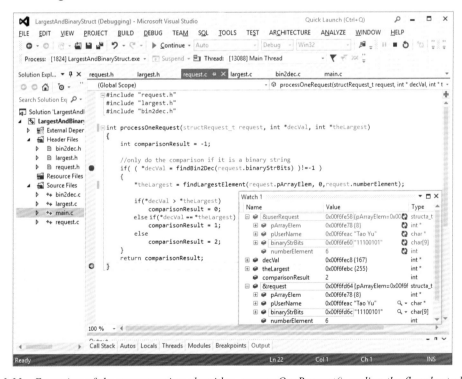

Figure 3.33   Execution of the programming algorithm, processOneRequest(), coding the flowchart algorithm shown on Figure 3.29

Figure 3.34 shows the end of the program execution where the statement in the body of "*case 2:*" of the switch statement is performed since the return value from *processOneRequest()* is 2 (i.e., *comparisonResult = 2*).

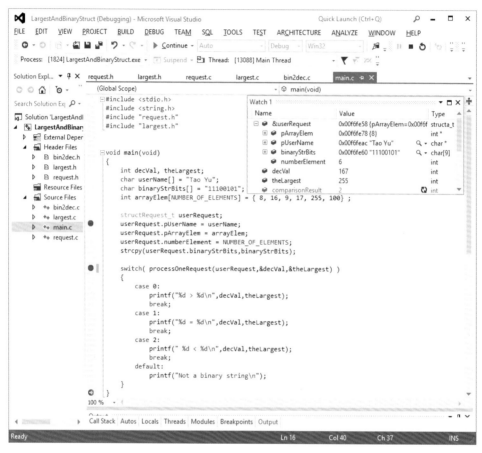

Figure 3.34 Concluding the execution of the new refactored main(), i.e., till third breakpoint after executing "case 2:"

Could you guess how structure variable, userRequest, is passed to processOneRequest() at the call site in Figure 3.34? Looking at red-marked addresses of *userRequest* variable (i.e., 0x00F6FE58) at caller-side and *request* variable (i.e., 0x00F6FD64) at callee-side on Figure 3.33, you will see they are different, which means the latter structure variable, *request,* is a copy of the former one, *userRequest*. Hence, the compiler-generated code was based on the passing-by value mechanism. Looking once again at blue-marked addresses (i.e., 0x00F6FE60 and 0x00F6FDC) on Figure 3.33 you will see that consequently the string field *binaryStrBits* is also passed by value. So, to pass an array by value, instead of the default by reference, you must declare it as a field of the structure and then pass the structure variable. For the effect, check the declared structure type, *structRequest_t,* on Figure 3.30 and also notice the reserved byte for '\0' given by the "+1" to avoid *strcpy()* call on Figure 3.34 from overwriting the next adjacent field, *numberElement*.

A more generic interpretation and implementation of the previous problem statement should be able to process more than one request from more than one user. The easiest way that will impact less the already written-code and let for you to practice, should be refactoring the *main()* to firstly populate an array of converting-comparison structure elements and secondly call in a loop *processOneRequest()* from the first until the last element. However, let's follow another approach to exercise the return of a structure variable from a function, while dealing with some possible issues (e.g., returning address of local variable and accessing a NULL pointer).

Figure 3.35 presents the program algorithm for populating the array of structure variable, *allRequests*, with two requests and then processing both on a loop, by successively calling of *processOneRequest()*. After each call, conversion-comparison results are registered in new fields of the previously populated *allRequests* variable, in their corresponding array entry, according to the request under processing. Finally, the filled *allRequests* constant pointer is returned from the callee, *processAllRequest()*, to its caller, *main()*, which programming algorithm is depicted in Figure 3.36. Looking at both figures, you will see the memory layout of the shared structure array by caller and callee, under their corresponding stack frame.

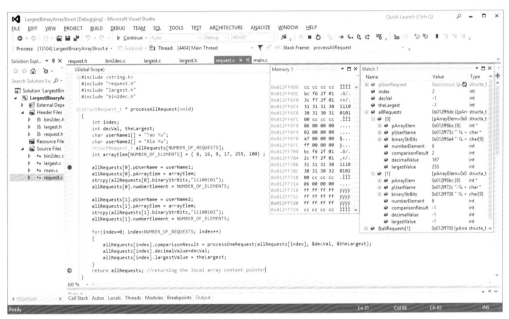

Figure 3.35 Execution of the programming algorithm for processing all request, which creates and returns a local structure array pointer

Figure 3.37 and Figure 3.38 show other two execution points of the *main()*, while registering the impact on the state of the pointer structure variable, *pUserRequest*, on calling *printf()* twice at different switch case selections. Listing 3.1 portrays the modified structure type, *structRequest_t*, to accommodate the last three integer fields, *comparisonResult*, *decimalValue* and *largestValue*. As said before, these fields are for conversion-comparison purpose and they are analyzed by *main()*, after being registered by *processAllRequest()*.

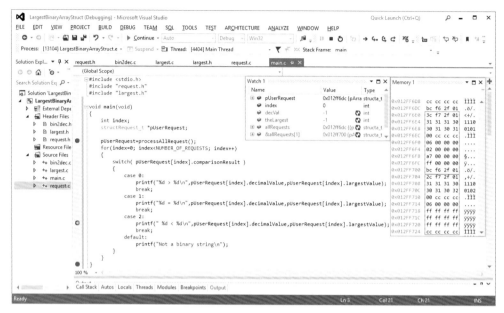

Figure 3.36   Execution of main() till the second break point and before the execution of the statement in the body of "case 2:" in order to register the state of the returned pointer structure variable, pUserRequest

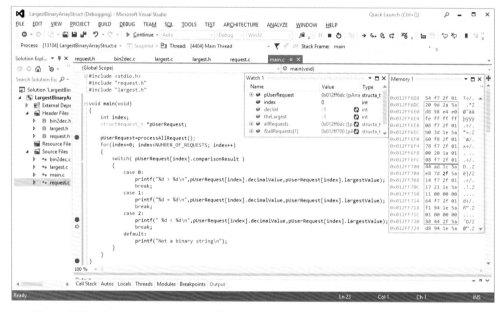

Figure 3.37   Overwritten impact on pUserRequest's state due to the printf() call at "case 2:"

As it can be seen from Figure 3.38, the execution of the program is concluded like everything is fine. But, *what about trying to access fields of pointer structure variable, pUserRequest, once again just after leaving the loop and before exiting from main()?*

3 HANDS-ON-POINTERS:THE BASIC   137

Figure 3.38   Overwritten impact on pUserRequest's state due to the printf() call at default selection. Notice, pArrayElem, the first field of the structure variable, pUserRequest, is now pointing to NULL (i.e., at the address 0x012FF6DC)

Listing 3.1   Header file for the request processing module, whose structure type was changed to accommodate conversion-comparison results

---

request.h

#ifndef REQUEST_H_

#define REQUEST_H_

#define NUMBER_OF_BITS   8
#define NUMBER_OF_REQUESTS   2

typedef struct structReq_tag
{
   int *pArrayElem;
   char *pUserName;
   char binaryStrBits[NUMBER_OF_BITS+1];
   int numberElement;
   int comparisonResult;
   int decimalValue;
   int largestValue;
} structRequest_t;

structRequest_t * processAllRequest(void);
int processOneRequest(structRequest_t,int *decVal, int *theLargest);

#endif REQUEST_H_

Being fully satisfied just because the program is executing as expected, is not a good practice, mainly when you are dealing with array of structure and pointers, as well as ignoring warnings prompted to you by the compiler, as seen in Figure 3.39. That is why this book goes against the printf-based dummy-kind of debugging and toward approaching the debugger as the driver or enabler of the C programming learning and teaching process. Leveraging debugging process at the core of C programming learning and teaching process, possibly undetected and latent runtime errors will be caught at the outset, avoiding costly future maintenance issues.

Let's then progressively fire those runtime errors by isolating different effects of the called *printf()* at switch case selections, as illustrated Figure 3.37 and Figure 3.38. To isolate the impact of the *printf()* called at the "*case 2:*" switch selection, let's firstly process only the first array entry (i.e., *pUserRequest[0]*) and secondly execute the blue-marked statements' block to access again structure fields (see lower blue-marked code on Figure 3.40). The programming algorithm for *main()* on Figure 3.41 using "*point-to*" operator instead of indexing operator, is exactly the same, in terms of behavior, as the one previously discussed and presented in Figure 3.37. Figure 3.41 shows the read trash data, while trying to access again the return pointer structure variable.

To trigger the runtime error due to the calling of *printf()* at the default selection, just remove the −1 from the upper blue-marked block on Figure 3.40 in order to also access the second entry of the pointer structure variable (i.e., accessing *pUserRequest[0] followed by pUserRequest[1]*). Rewinding back to Chapter 2 and specifically to Figure 2.31, to remember that the address 0 is used to define the NULL pointer which belongs to kernel space, instead of user space, and thus, its access is forbidden from a user-defined program. That is why the second execution of the program to access both entries after calling both *printf()*, will result in the crash illustrated by Figure 3.42.

Figure 3.39 Ignored warning regarding "returning address of local variable" that later manifested as runtime error

Figure 3.40 Execution with no crash but outputting trash data while accessing the pointer structure variable, pUserRequest, after calling printf() at "case 2:". The first blue-marked block shows the isolation of the impact associated to called printf()

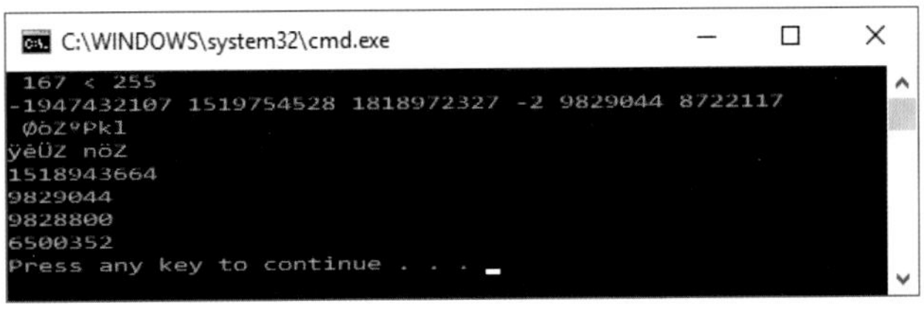

Figure 3.41 Displaying trash data overwritten on the array of structure by calling printf() at "case 2:"

Now comes another question: *why is returning the address of local variable so dangerous?* Once again, you should rewind back to Chapter 2 and specifically to variable scope and lifetime. Since the structure array variable, *allRequests*, is local to *processAllRequest()*, its lifetime will be over as soon as *processAllRequest()* returns, and thus, its memory location on the stack becomes free for any further usage (e.g., for the assigned stack frames of the called *printf()*). *How can this problem be solved, based only on the knowledge you have acquired so far?* The solution consists in changing the lifetime of all pointer variables, locally declared in the body of *processAllRequest()*, for the duration of the program execution, by preceding their declarations with static storage qualifier (see the blue-marked block in Figure 3.43). Check the output window during the compilation time to see that the warning regarding "*returning address of local variable*" disappeared. Since these local pointers are no longer allocated on the stack due to the static qualifier, the state of the *allRequests* variable is no longer overwritten by arguments passed to *printf()*, at the two discussed call sites (see Figure 3.43 to Figure 3.45).

Figure 3.42   Program crashing due to the impact of calling printf() at the default switch selection, which overwritten the pArrayElem field to point at a NULL pointer (see Figure 3.38)

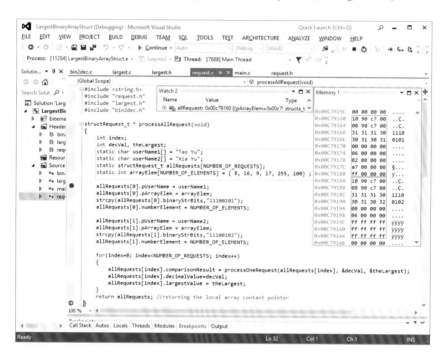

Figure 3.43   Execution of processAllRequest() with all local pointers declared as static, meaning they will be no longer allocated on the stack segment. They will be allocated on the data segment

Figure 3.44 and Figure 3.45 also illustrate pointer arithmetic on the pointer structure array, *pUserRequest*. Looking at the *Watch window*, and knowing the pointer to *pUserRequest[0]* which is 0x00C79160, you will see the pointer structure variable, *pTemp*, evolving successively from 0x00C79160 to 0x00C79184 to 0x00C791a8. The scale factor is 0x24=36 bytes, exactly the size of *structRequest_t* (i.e., 0x00C79184 – 0x00C79160). Finally, Figure 3.46 shows a clean execution of the program after fixing the previous warning and keeping the state of the pointer structure array, *pUserRequest*, unmodifiable from any execution side-effect.

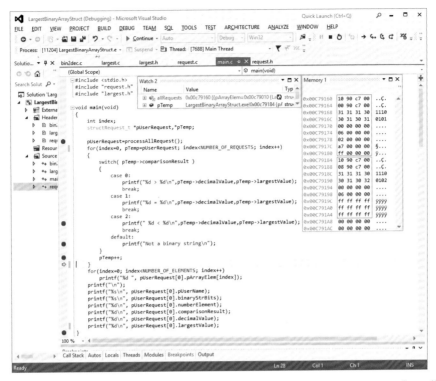

Figure 3.44   *The state of the pointer structure array, pUserRequest, is no longer overwritten after calling printf() at the "case 2:" call site*

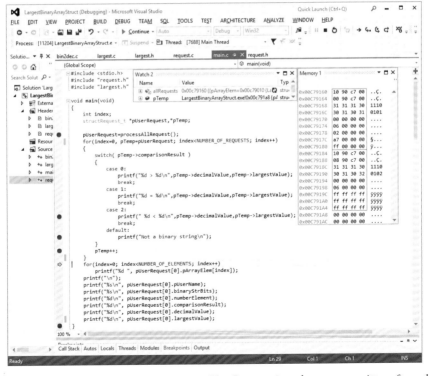

Figure 3.45   *The state of the pointer structure array, pUserRequest, is no longer overwritten after calling printf() at the default selection call site*

Figure 3.46   Clean results at the end of the program execution after changing the lifetime of all local pointers by qualifying them as static storage

To better consolidate the knowledge introduced in this section, please explore and discover the impact, if any, of packing mechanism on pointer arithmetic.

This chapter main focus were on understanding and practicing the basics of pointers and it will end now with the following recommendations:

Recommendation 16: Always initialize pointers at both pointer- and pointee-levels before use them because uninitialized pointers are particularly risky, resulting in hard-to-find and serious bugs. Trying to access a NULL pointer will crash a program, as shown in Figure 3.42.

Recommendation 17: Always use *constant* reference when passing-by reference large data objects (e.g., large arrays or structures), unless you need to change the value of the argument (a complement to Recommendation 11).

Recommendation 18: Although *nibble* was used to name the formal parameter when declaring *isBinary()*, always assigned name to pointer variables starting with lowercase *p* followed by a noun (e.g., *pNibble*) to make program understanding easier.

Recommendation 19: Always pass and return structures to functions by-reference and access them indirectly through pointers for efficiency purpose, mainly if they are large (a complement to Recommendation 11 and 16). Recall C language does not support reference parameters automatically and thus, by default all parameters, except array, are passed by value.

Recommendation 20: Although you can strictly order structure fields to minimize padding, it also harm code readability. As programs are not only communications to a computer but also communications to other people, whenever possible order structure fields in semantically coherent groups with related pieces of data kept close together. In doing so, the design of structures will communicate the design of a program.

Recommendation 21: Although performance issues are not a concern by now, beware that intelligently reordering of structure fields by taking into account the program's data-access pattern can leverage better cache locality. By intelligently reordering, we mean grouping related and also co-accessed data in adjacent fields to fit them within a cache line, which usually begins on a self-aligned address (a complement to Recommendation 20).

Recommendation 22: Do not simply ignore warnings as some of them can later manifest as huge runtime errors, which will be sometime undetected and dormant until the deployment of future version of the program.

References

[9] HORTON IVOR, Beginning C. 5th ed. New York: Apress, 2013.

[10] REESE RICHARD, Understanding and Using C Pointers. 1th ed. Sebastopol: California O'Reilly Media, 2013.

Several other information sources were also used, mainly from internet, as well as those previously referenced in previous chapters. Thus, the credits also go to all them.

4 HANDS-ON-POINTERS: ADVANCED FEATURES

Learning objectives

1. Understanding and practicing dynamic memory management in C.
2. Understanding and working with file pointers.
3. Understanding and practicing with pointers to functions.
4. Understanding and practicing passing arguments to main().
5. Practicing the pointer' usage by examples.

Theoretical contents

1. Pointers and functions.
2. Command line arguments: arguments to main().
3. File system basics and the file pointer.
4. Dynamic memory management in C and Linked list.

Strategies and activities

1. Always exercising each presented topics by first exemplifying along each theoretical introduction and then presenting right away similar problem to be completely and individually solved by students.
   a. Always offers bonus to the first students finishing the exercise as well as to those presenting best solutions.
   b. Bonus should be also offered during all classes starting from the first one.
   c. Choose some of the students with best solutions for briefly presenting their solutions and also helping other students altogether with the instructors.
   d. Give bonus to students revealing certain level of intellectual curiosity on a sequence of two classes, at least.
2. Running a simple quiz about all pointer' knowledge applied and covered in this chapter.
3. Ending with a project made by group of students, tackling full C program development lifecycle, supported by a report describing each phase, a short presentation and a final discussion.

Apart from regular pointer variables used so far to deal with static allocated variables, there are others special purpose pointers and usages, such as file pointers, function pointers, command-line pointer arguments and pointers to dynamic allocated data. Since the main purpose of a computer system is creating, processing, storing, searching and retrieving data, the file system became an essential OS component in solving the problem of storing and organizing information on media, as it provides the machinery to support such operations. A filesystem can be described as an OS process in which files and directories, represented by file pointers, are named, protected and logically located for storage and retrieval, and thus abstracting the user from the internals of the permanent storage devices, such as hard disk drive, CD-ROM or DVD. Not all existing programs are meant to communicate with users, such as those that are only invoked by the OS. The only way to input data to these OS services, is through command line pointer arguments. Furthermore, it makes programs scriptable as scripts or other programs can call a program and pass it data to operate on. For instance, by double-clicking a file name or icon in Windows OS, that file's name will get passed by the OS as an argument to the callee, i.e., the file system opening service. Function pointers are extremely efficient and flexible in creating callback mechanism where execution logic are passed as arguments to functions, resulting in a more customizable code during execution. It can also be used to invoke plugin functions, implemented as a dynamic-linked library loaded at runtime. Dynamic memory allocation is essential when is not possible to establish a worse-case requirement for memory usage, making impossible the static allocation of all needed memory, without severe memory waste.

4.1 Pointers and Functions

Previously, programmer and machine views of all program objects were discussed and expressed through an externally 5-tuple and internally 2-tuple attributes, respectively. The external 5-tuple can be extended to a generic 7-tuple to accommodate storage class attributes, as both code and data come also with their corresponding scope and lifetime. Based on the internal (location, content) tuple, pointer to function or function pointer is nothing else than a variable which points instead at a code object at a given location/address and whose content/pointee specifies some behavior[①] instead of value. The name of a function as presented in a function definition is also mapped by the compiler to a constant pointer, serving later as: (1) the entry point of that function when it is called, (2) an argument passed to caller functions and (3) a pointer variable returned from another function.

Pointers to functions are extremely efficient and flexible in supporting runtime customization, as they can easily promote alternate execution paths to replace selection statements, while realizing late-binding. Late or dynamic binding is a computer programming mechanism

---

[①] Semantically, compilers and processors interpret differently the pointee a function pointer points at, while allocating it to read-only memory segment.

which defers, till the runtime, the choice regarding the proper function to be invoked. For example, the programmer can decide at runtime for executing functions in a sequence which was unknown at compile-time and with no use of control constructs statements. However, although not always guaranteed, the use of function pointers can cause some slowdown as it can inhibit jointly operation of some modern processor internal circuits such as branch predictor[①] and pipeline[②].

Putting side-by-side the following generic function prototype and the generic function pointer declaration, one can conclude they look very similar:

1. return_type functionName (parameter_list);
2. return_type (*p2F) (parameter_list);

Removing the set of parentheses on the left-side from the function pointer declaration in point 2, it will become like the function prototype in point 1, but returning instead a pointer. It is exactly the parentheses that make it a pointer to a function with the name of *p2F*. *How to generically read the above function pointer declaration?* Following the "right-left" rule:

1. *p2F* is a funciton.

Step 1: Find the identifier

2. *p2F* is a function **receiving parameter_list.**

Step 2: Look at the symbols on the right of the identifier

It is *(parameter_list)* which indicates it as a function.

3. *p2F* is a **pointer to** function receiving *parameter_list*.

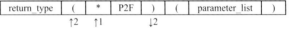

Step 3: Look at the symbols on the left of the identifier

It is the dereferenced value of p2F, i.e. (*p2F) which indicates it as a pointer to function.

4. *p2F* is a pointer to function receiving *parameter_list* and returning return_type.

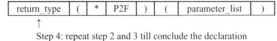

Step 4: repeat step 2 and 3 till conclude the declaration

To simplify the use of function pointers to novices, it is convenient to declare a type definition for function pointers, using the *typedef* keyword:

    typedef return_type (*p2F_t) (parameter_list);

As claimed above regarding a flawless usage of data object pointers, always follow exactly the same 3-step approach (i.e., declaration, initialization and finally use of a pointer) with function code object pointers (i.e., function pointers). That is, function pointers can be declared, assigned values and only then used to access/run the functions they point to.

---

① Branch predictor is a digital circuit that comes with modern processor to guess beforehand which is the next execution path when facing multiple execution sequences, as dictated by control constructs.
② Pipeline circuitry improves processor performance by overlapping execution of multiple instructions.

4 HANDS-ON-POINTERS: ADVANCED FEATURES • 147

To exercise the above theoretical introduction to function pointers let's imagine one more demand comes from users of previous program, to allow them choosing between different alternate implementations of *bin2dec()* at setup time, according to their preference. Maybe, they will even ask for changing such algorithm dynamically at any execution instant. The new revised problem statement will be: "*Compare a binary number, represented as a binary string, to the largest number found in a sequence of numbers, as required by a user who can specify his preferred conversion algorithm*".

For the maintenance purpose, an addenda with the following items or points will be discussed and added to the project:

1. The choice of the conversion algorithm will happen only at program setup, but we recommend you to extend it later to enable the user to change it to the one that better fits to his need, at any given instant;
2. Somehow, the preferred or selected conversion algorithm should be registered, after being chosen from a list of options;
3. Alternate binary to decimal converters will be added to the list of options, for example, by devising another one that is able to process the binary string in the reverse order of power of two;
4. Two binary to decimal conversion functions will be registered by their pointers in an array of function pointer, while the previous *structRequest_t* will be extended to include a callback (i.e., a function pointer to the chosen converter);
5. Functionalities previously offered in the *request* module will be accordingly refactored to allow the selection of the converter by users.

To support the late-binding mechanism required in the revised *request* module, a header file illustrated in Listing 4.1 is included in the project with the definition of a function pointer type according to the prototype of binary string converter programming algorithms, such as *findBin2Dec()* and *findBin2DecReverse()*. This header file is later included in the revised *request.h* shown in Listing 4.2.

Listing 4.1 Extra header file to support late-binding required by the revised request processing module

latebinding.h

```
#ifndef LATEBINDING_H_

    #include <stddef.h>          //for NULL definition

    #define SETUP_CHOICES   2

    typedef int (* pFBin2Dec_t)(char * const);

#endif LATEBINDING_H_
```

The revised *request.h* extends *structRequest_t* with a field, *p2Fbin2dec*, to accommodate the chosen function pointer for binary to decimal conversion function, using the function pointer type, *pFBin2Dec_t*, previously defined in "*latebinding.h*". It also presents the revised prototype for the new *processAllRequest()*, refactored to dynamically accommodate the selection of the binary converter function. It presents as formal parameter a callback table, *cbkArray[]*, with function pointers to all alternate binary converter functions alongside a brief description of them, for building a simple text-driven menu. As an exercise, you should later define a structure data type to merge both type of data as individual fields.

Listing 4.2   Revised header file for the request processing module with late binding support

```
                                    request.h
#ifndef REQUEST_H_

  #define REQUEST_H_

  #include "latebinding.h"

  #define NUMBER_OF_BITS          8
  #define NUMBER_OF_REQUESTS      2

  typedef struct structRequest_tag
  {
    char binaryStrBits[NUMBER_OF_BITS+1];
    int *pArrayElem;
    char *pUserName;
    pFBin2Dec_t p2Fbin2dec;  //bin2dec callback
    int numberElement;
    int comparisonResult;
    int decimalValue;
    int largestValue;
  } structRequest_t;

  structRequest_t * processAllRequest(pFBin2Dec_t cbkArray[], char **purposeArray);

#endif REQUEST_H_
```

Figure 4.1 presents the programming algorithm for *main()* and its partial execution before entering the refactored *processAllRequest()*, by passing arguments according to its new prototype with formal parameters related to callback table and description (see the lower blue-marked statement block). Firstly, this table is globally created and fully initialized with NULL function pointers (see the upper blue-marked statements block). Secondly, it is filled in with function pointer to alternate binary converters, *findBin2Dec()* and *findBin2DecReverse()*, as shown by the middle blue-marked statements block. The middle blue-marked statements block also presents two alternate ways to get the constant address of a function, by simply naming it. Although optional to most compilers, to write portable code, it is convenient to prefix the function name with the "address-of" (&) operator.

Still in Figure 4.1, the upper red-marked blocks on the *Watch window* shows the callback table first entry, *callbackArray[0]*, after being populated with a real non-NULL function pointer, while the *Memory* window presents the pointee at that address.

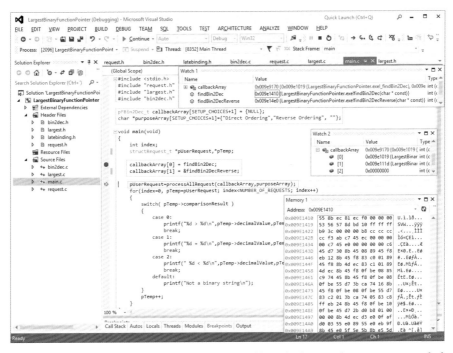

Figure 4.1   Entering the main() by initializing the callback table with alternate binary converters, before starting request processing by call the refactored processAllRequest()

Figure 4.2 exhibits partial execution of *processAllRequest()*, by exercising two ways to choose at setup-time the user's preferred binary converter function, among those available in the callback table (see the two call sites depicted by blue-marked statements). According to the execution point, as indicated by the yellow arrow and the NULL callback of the *allRequest[1]* as shown by the red-marked blocks, only the first call site was reached. The callback chosen by the user was *findBin2DecReverse()*.

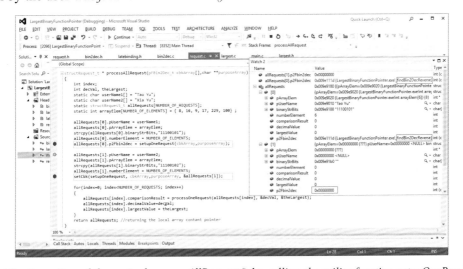

Figure 4.2   Execution of the revised processAllRequest() by calling the utility function setupOneRequest() to choose and set the callback for the first request, i.e., allResquests[0].p2Fbin2dec

Basically, to leverage the dynamic selection of the preferred binary converter, the *processAllRequest()* was refactored with the support of the two utility functions at the call sites, whose programming algorithms are exhibited in Figure 4.3. Algorithmically, *setCbk()* and *setupOneRequest()* are exactly the same, *as* the former is just a wrapper function[①] of the latter. The *setupOneRequest()* presents the user with a text-driven menu regarding alternate binary converters and wait until a valid choice is performed, before returning the chosen function pointer back to the caller, *processAllRequest()*.

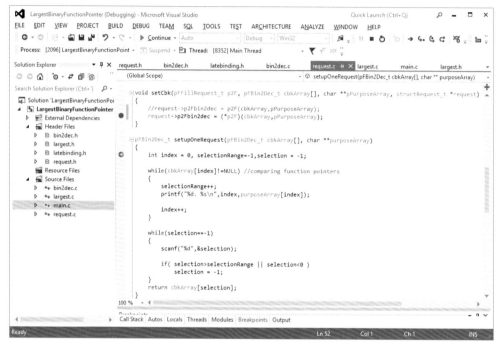

Figure 4.3   Utility functions for supporting the offered late-binding for binary converter setup at runtime. Notice that setCbk() is just a wrapper of setupOneRequest()

Programmatically, *setupOneRequest()* receives as arguments an array of pointer to functions and an array of pointer to strings which shortly describes used conversion approaches, while returning a function pointer to the selected binary converter. The rightmost formal parameter, *purposeArray*, given by a pointer of pointer to char (i.e., *char **purposeArray*), represents also a one-dimensional array of string pointers. Two levels of indirection or two pointee-levels will be required to fully dereference it. As an exercise to consolidate your knowledge, you should run a comparison among variable declarations based on *char **p*, *char *[]* and *char [M][N]*, to explore their memory layouts and discover their similarities.

According to the execution point as indicated by the yellow arrow on Figure 4.4, *setCbk()* was executed at the second call site and the chosen callback by the user was *findBin2Dec()*. The first formal parameter of *setCbk()* is "a pointer to a function that returns a function pointer to a different function," as shown by the upper blue-marked statement. Such formal parameter is compatible with the prototype of *setupOneRequest()* and thus, used to pass-

---

① A wrapper function is a function whose main purpose is to call another function with little or no more extra statements.

by reference a pointer to this latter function as argument. Back to Figure 4.3, you will see why *setCbk()* is a wrapper of *setupOneRequest()*—its body has only one statement, exactly to invoke *setupOneRequest()* through the previously passed function pointer. Two calling alternatives are presented, one commented and the currently used, which dereferences the function pointer before calling the function, *setupOneRequest()*, it points at. This explicitly dereferenced calling approach used to be the preferred one, to clearly distinguish this kind of call-by address from the regular call-by name.

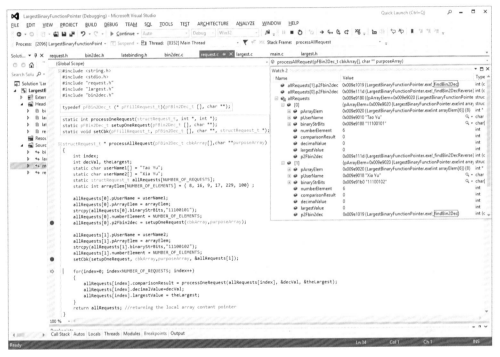

Figure 4.4 Execution of the revised processAllRequest() by calling the utility function setCbk() to choose and set the callback for the second request, i.e., allResquests[1].p2Fbin2dec

To conclude the refactoring of the *request* module, there are still two remaining activities to be performed:

1. The *processOneRequest()* is no longer called outside the request module and its scope as well as those of *setupOneRequest()* and *setCbk()* will be restricted to *request.c*, using the static storage class qualifier (see the lower blue-marked statements block in Figure 4.4);
2. The *processOneRequest()* is slightly modified to invoke the binary converter functions e.g., *findBin2DecReverse()* and *findBin2Dec()*, through their corresponding function pointers, as shown in Figure 4.5.

The blue-marked statement block in Figure 4.5 illustrates the usage of the alternate calling approach that is very similar to the calling of a function by-name, i.e., without dereferencing the function pointer.

Figure 4.6 presents an alternate binary string conversion to decimal value, by reversing the corresponding power of two associated to each bit according to the regular bit position in a byte. Comparatively to the older version, *findBin2Dec()*, that has been used from the very beginning, only the indexed power of two was modified by assigning the higher exponent to the first bit, instead of zero as previously.

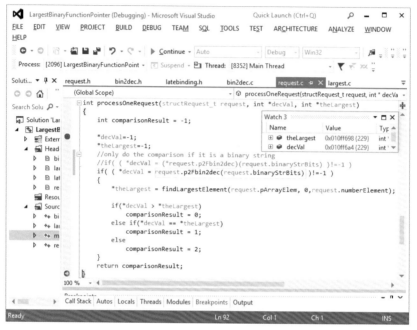

Figure 4.5 Execution of the revised processOneRequest() with constrained scope and calling of the binary converter through function pointers

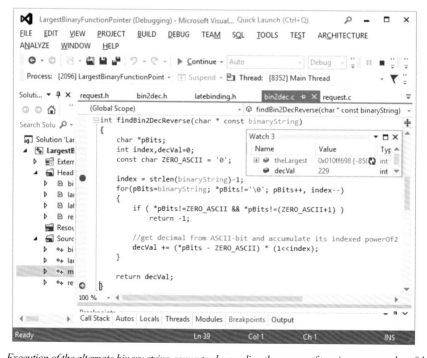

Figure 4.6 Execution of the alternate binary string converter by grading the power of two in reverse order of the bit layout

Finally, in Figure 4.7 it is shown the final result of the program execution altogether with the selected users choices for binary converter callback of each request. The red-marked block illustrates the two execution of *setupOneRequest()*: a menu is displayed with possible options and the user chooses one of them. The blue-marked block illustrates the result after processing the two previously configured requests, as performed by the *for-loop* in the *main()* body. It shows that one of the inputted requests has an invalid binary number. Furthermore, as *setupOneRequest()* forces the user to always choose only among non-NULL function pointers, it enables blind call of callback pointers in *processOneRequest()* without any further checking (i.e., if they are guaranteed to be not NULL at the call site). However, without this kind of previous validation, you always need to check pointers before using them (e.g., "*if ((p2Fbin2dec && *decVal = request.p2Fbin2dec(request.binaryStrBits))!=-1)*").

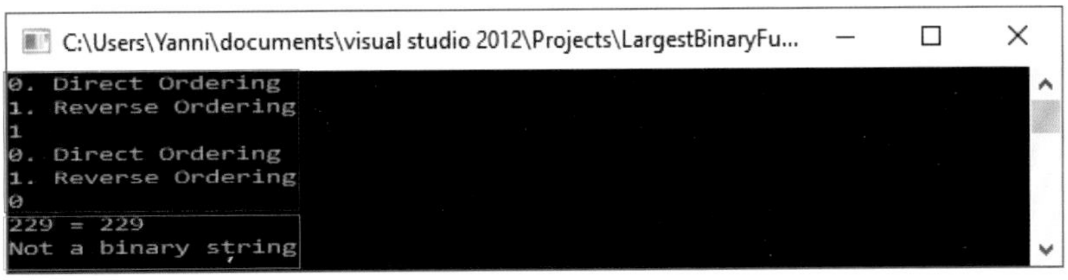

Figure 4.7   Illustrating the configuration at setup-time and the final result at the end of execution

4.2 Command Line Arguments: Arguments to Main()

In Chapter 1, the structure of a C program was introduced in Listing 1 as consisting of a set of declarations and functions welded by the special and mandatory *main()*, where execution starts and which is in charge of directly or indirectly calling other functions. For example, in Figure 4.1 *main()* calls *processAllRequest()* which in turn calls *processOneRequest()* and so on. Looking back to the Listing 1.1, you will realized *main()* was presented through a generic function prototype of "*return-type main(parameter_list);*", while we have been using so far a specific prototype given by "*void main(void);*".

Furthermore, knowing the ANSI C standard endorses "*int main(void);*" as a prototype, maybe you are feeling confused and wondering how many specific prototypes exist and are allowed for *main()*, as well as which one should be used. Well, there are several issues still open to debate about this matter and it is guaranteed that the range of valid *main()* prototypes is much larger. However, before going into further explanations, let's explore what really happens when returning from *main()* and then try to answer the following questions based on two execution scenarios:

1. What is the *main()* caller?
2. What does *main()* return?
3. How is the return from *main()* different from a regular function return?
4. Why there are several prototypes for *main()*?
5. When should a given prototype be used?

For the first scenario, the previous *main()* used so far (i.e., with the "*void main(void)*" prototype) will be debugged by removing all previously set breakpoints, except the one at the closing bracket, as showing in Figure 4.8.

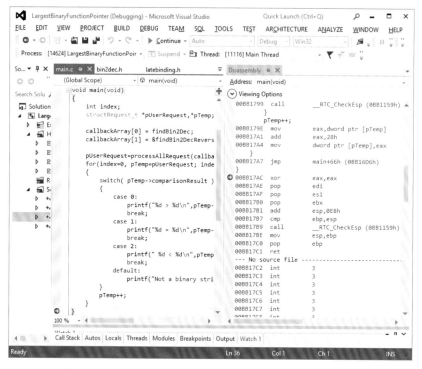

Figure 4.8 Debugging the return from main() with "void main(void);" prototype: placing a breakpoint at the last closing bracket of the main() body

The following observed facts can be numerated:

1. Internally a *main()* returning an *int* and three formal parameters was used instead, as shown by the blue-marked block of statements in Figure 4.9;
2. Even a user-defined with a return type of *void*, internally *main()* is still returning an *int* through the *mainret* variable (see the blue-marked block of statements), whose value is set to zero (see the red-marked block on the *Watch* window), as shown in Figure 4.9;
3. Since *argc* argument is equal to 1, it means no argument is passed by the OS, and thus, the *argv* argument carries only the full path name of the executable program console application, as shown by the two red-marked blocks in the *Watch* window of Figure 4.9;
4. The closing bracket at the set breakpoint is effectively converted into an *exit()* operating system (OS) call, as shown by the blue-marked block of statements on Figure 4.10.

Based on the observed facts, the first 3 questions can be easily and right away answered as: (1) the range of valid *main()* prototypes is much larger than those we have being discussed so far, (2) the caller of the *main()* is the OS and thus, the return from the initial[1] call to *main()*, when reaching the closing bracket, is converted to an *exit()* system call[2] which registers with

---

[1] There is nothing wrong from the ANSI C standard point of view with calling *main()* from a program. However, we do not recommend it, even though any other call to *main()*, besides the initial one, behaviors as a regular function (i.e., it returns regularly since its caller will be a regular function, instead of the OS).

[2] A system call or kernel call is a programmatic way to provide interface to OS services. For instance, a user-written program can use a system call to request the opening of a file.

zero a successful or normal program termination.

Figure 4.9 Debugging the return from main() with "void main(void);" prototype: showing the MS Visual C compiler implementation-defined main() internal prototype which returns an int and receives 3 arguments.

Figure 4.10 Debugging the return from main() with "void main(void);" prototype: converting the closing bracket into a call to exit system call to register the state of program execution and terminate the program corresponding OS process

For the second scenarios, the *main()* will be slightly changed to return an *int* and it will be also appended a "*return 0;*" statement before the closing bracket, where a new breakpoint is also set (see Figure 4.11).

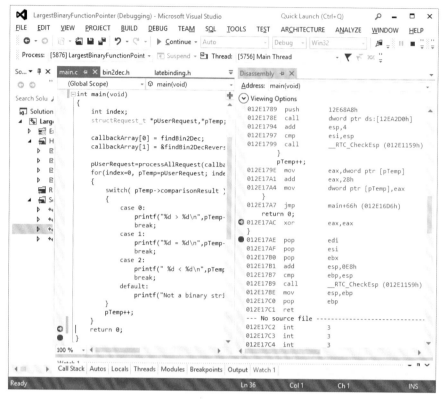

Figure 4.11   Debugging the return from main() with "int main(void);" prototype: placing a breakpoint at the return statement

Repeating previous debugging steps, you will see that after reaching and executing the "*return 0;*" statement, the observed behaviors are exactly those ones of the previous debugging scenario, as shown in Figure 4.11 to Figure 4.13. Since the ANSI C standard endorses *int* as the return type, the MS Visual C compiler follows the standard conforming implementation which by default return an *int*, no matter the used prototype. Try to use "*main(void)*" or "*main()*" prototypes to see that the MS Visual C compiler will always follow the same conforming implementation. Reading the C11 document (i.e., ANSI C standard), it refers to two forms of conforming implementation for execution environments: (1) the hosted environment like the one used in this book, which is based on Windows OS and (2) the freestanding or bare-metal environment, usually used when designing your own operating system or low-end embedded system firmware. The appended "*return 0;*" statement in Figure 4.12 seems to be unnecessary and so, it can simply be omitted for two reasons. Firstly, it might confuse novice programmer and secondly, as you saw above, a compiler following ANSI C conforming implementation will automatically generates it at the end of *main()*, if no other statement is specified. Although either practice is acceptable, try to be consistent or simply do not rely on the standard implementation-conformity of the in-use compiler, if strictly portability among confirming implementations is a concern.

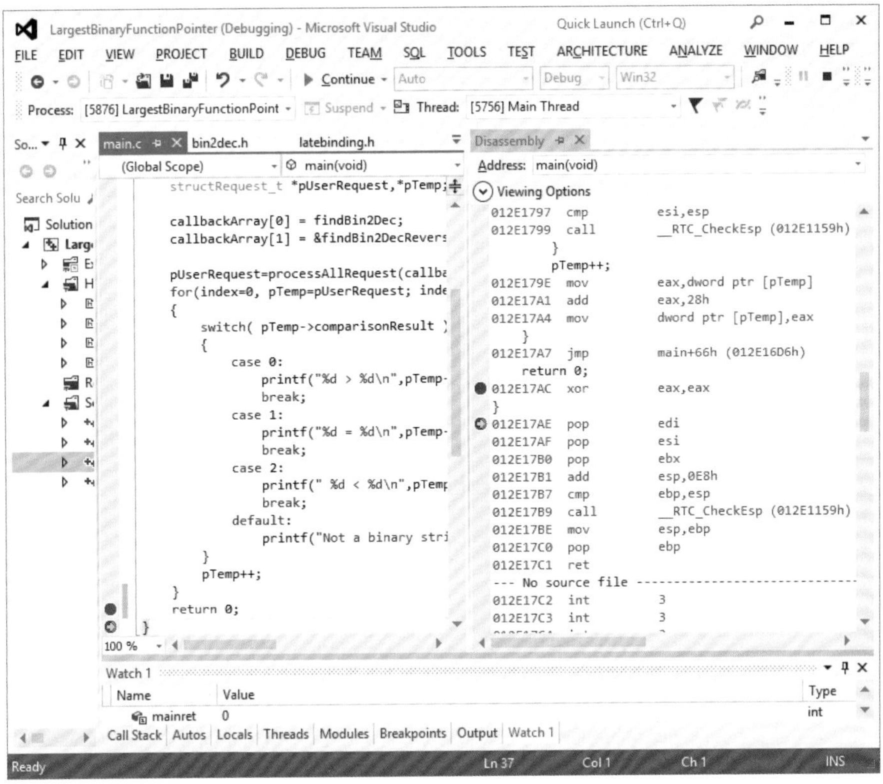

Figure 4.12   Debugging the return from main() with "int main(void);" prototype: after executing the return statement

Does the standard recommend any implementation-conforming for the return type under bare-metal execution environment? For instance, for backward compatibility with traditional C (a.k.a., K&R C [1]), even "main(void)" prototype is supported by a hosted standard conforming implementation compiler, while in a freestanding environment the global effect of program termination is implementation-defined. That is, the generated code is based on implementation-defined behavior undefined by the standard. The standard also defines two strictly conforming (i.e., not implementation-defined) values for returning, EXIT_SUCCESS and EXIT_FAILURE, as provided in *stdlib.h*. However, a bare-metal program may need an implementation-defined manner to discriminate among several unsuccessful terminations. [1]

Is the main() prototype receiving a 3rd argument, envp, as shown in Figure 4.9 and Figure 4.13, strictly portable among standard conforming implementations? Checking again the C11 document, it only explicitly mentions the following two prototypes: "*int main(void)*" and "*int main(int argc, char *argv[])*", being the latter equivalent to "*int main(int argc, char **argv)*". The prototype given by "*int main(int argc, char *argv[], char *envp[])*" is valid under some OS (e.g., Windows and some Unix-flavour like Linux). MS Visual C and GNU GCC implementations permit it, making a program that uses it implementation-defined and no longer maximally portable among conforming implementations. The 3 formal parameters have the following meanings:

---

[1]   K&R C is the original C language dialect introduced by the Brian Kernighan and Dennis Ritchie book entitled "*The C Programming Language*".

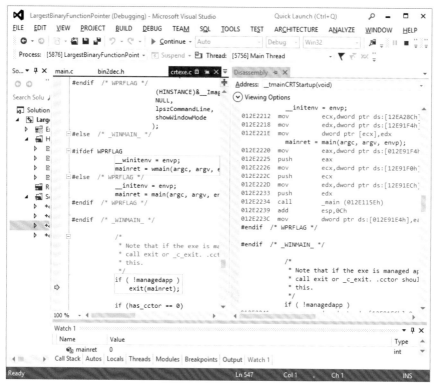

Figure 4.13   Debugging the return from main() with "int main(void);" prototype: converting the closing bracket into a call to exit system call

1. *argc* is the argument count indicating how many arguments are passed by the OS to the program. At least one argument is always passed which is the name of the executable as shown by the left-side red-marked block in Figure 4.9, for a *main()* receiving no *parameter_list*.
2. *argv* is the argument vector viewed as a one-dimensional array of string pointers, where each pointer references a command line argument. The first string, *argv[0]*, will always contains the program name, as shown by the right-side red-marked block in Figure 4.9.
3. *envp* is the environment vector which contains environment variables. Environment or shell variables are a lightweight approach to communicate with a program when compared to the traditional configuration file approach, in which large amount of data is stored to be used by a program every time it is run.
4. *argc*, *argv* and *envp* are only traditional names. You may assign any other names to them.

To answer what to return from *main()* and which prototype to choose, we should say it depends on what you want to do with the executable program and if you want to change/specify the program configuration during setup, respectively. For instance, for the former case, you may need to write a batch file for Windows-based environment or shell scripting in Linux-based environment to examine the return code from any program. For the latter case, you may need to provide a new configuration any time a program is run.

Now the previous program can be changed to pass to the *main()* the indexes for the two function pointers corresponding to the two preferred binary converters. To specify command-line arguments for debugging follow the following steps:

1. Select the project in the *Solution Explorer* on the left-side of MS Visual studio screen;
2. Right-click the project and then click properties or alternatively click properties from the *Project* menu, and it will appear the following Property Pages on Figure 4.14;
3. Open the *Configuration Properties* folder, and then click *Debugging*;
4. In the *Command line arguments* field, enter the command-line arguments you wish to use, separated at least by one white space (see the red-marked boxes on Figure 4.14).

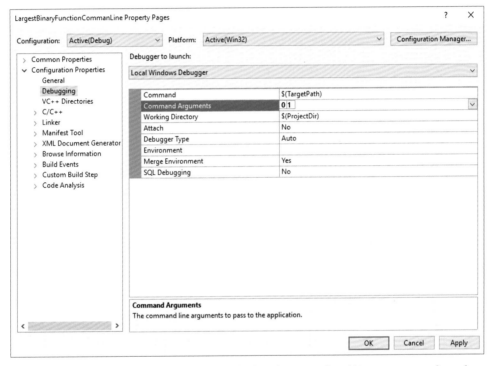

Figure 4.14 Inserting command-line arguments 0 and 1 for selecting preferred binary converters for each request

Figure 4.15 presents the refactored *main()* programming algorithm, as well as the initial state of the callback array and a completely invalid preferred callback choices for each request, as shown in the *Watch 2* window. First, it is necessary to include the prototype for the *exit()* system call and the constant returned value, *EXIT_FAILURE*, in case of abnormal program termination (see the uppermost blue-marked block in Figure 4.15), which will be later used by *check4ValidCbk()*. In the next blue-marked statements block, the callback table, *callbackArray*, is initially setup with the two available binary converters, while entries on a users' preferred choices table for binary converters, *preferredCbk*, are fully marked as invalid with *-1* indexes. The lowermost blue-marked statements block validates arguments received through command-line and shown in the *Watch 1* window, before calling the refactored *processAllRequest()*.

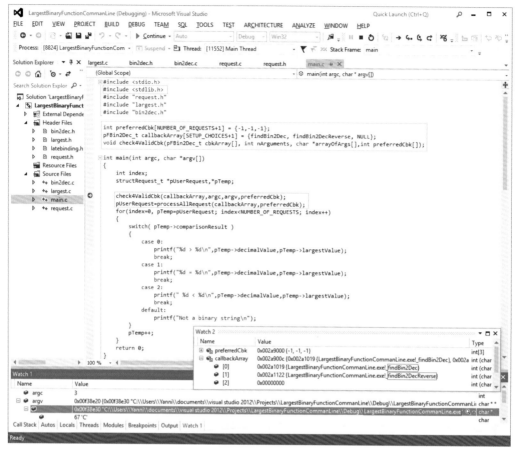

Figure 4.15 Partial execution of the refactored main() with new prototype for receiving command-line arguments, while calling a check4ValidCbk() to validate the inputted command-line arguments

Figure 4.16 presents the *check4ValidCbk()* which partially replaces *setCbk()* and *setupOneRequest()* of the previous program. Algorithmically, it checks for valid command-line arguments and only then it updates *preferredCbk* (see *preferredCbk*'s new state in *Watch 2* window) with user selections in terms of preferred binary converters. The middle blue-marked statement block shows how, for this specific case, string pointers as received from the OS can be converted to integer indexes. Otherwise, it aborts the program execution by calling the *exit(EXIT_FAILURE)* system call (see blue-marked statements blocks). Firstly, it checks if the number of passed command-line arguments is two as expected and secondly, if the inputted arguments will index valid entries on the callback table, *cbkArray*.

The refactoring of *processAllRequest()* followed the steps below in Figure 4.17:

1. Its declaration was changed by replacing "*char** purposeArray*" with "*const int preferredCbk[]*";
2. Previous call sites with *setupOneRequest()* and *setCbk()* were replaced with statements that will directly index the callback table, according to the selection setup on *preferredCbk*.

By concluding the debugging process, let's then execute the final executable program from *DOS* command-line window as shown by Figure 4.18. Notice that the name of the executable program is followed by two indices corresponding to the preferred binary converters for each request.

4 HANDS-ON-POINTERS: ADVANCED FEATURES • 161

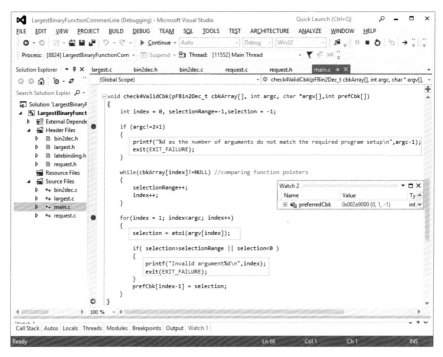

Figure 4.16   Execution state of the programming algorithm for check4ValidCbk(), which is in charge of validating the received command-line arguments

Figure 4.17   Execution state of programming algorithm for the refactored processAllRequest(), showing the assignment of binary converter callbacks to each request, according to the inputted command-line

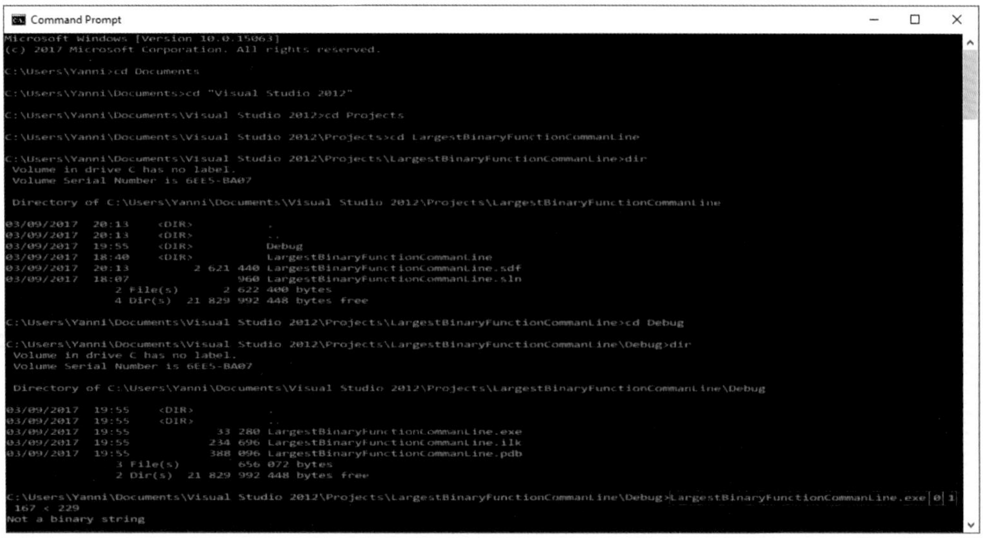

Figure 4.18   Executing the executable program from DOS command line by passing directly the two arguments 0 and 1

4.3   File System Basics and the File Pointer

Before introducing and using file pointers, let's firstly and briefly introduce you to what a file system is from both OS and programming language perspectives, what it manages, and what abstractions it provides to OSes and programmers. Secondly, you will see how the C file system (i.e., the C standard Input/Output) interoperates with the regular file system as leveraged by the OS.

To provide computer users with a uniform logical view of stored information, OSes usually come with the so-called file system, which is a mechanism for representing, storing, retrieving and organizing raw data in permanent or non-volatile storage devices (e.g., hard disk, CD ROM and pen driver). Therefore, any file system abstracts from the physical properties of storage devices by organizing its high-level information into structures, which typically include two fundamental abstractions or concepts of file and directory. Together, these two logical concepts create a file hierarchy system as commonly seen in Windows or Linux OSes. Among possible file system functionalities are tracking allocated and free space, maintaining directories and file names, and tracking where each file is physically stored on the storage device. The basic abstraction of file defines a logical storage unit for user data, consisting of metadata and the actual raw data, which are both recorded by OSes on permanent storage. The metadata component of a file describes properties or attributes about the raw data (e.g., the file name, the owner of a file, size, creation time, last modification time, type, location and access rights) a file systems needs to keep track of for a file, while the latter is the real content of a file. Naming a file makes it independent of the OS process, the user, and even the system that created it. Thus, while a user might create and assign a name to a file, other users might edit it by specifying its name. The owner can copy it to a pen driver or transfer it through the network. The directory is a file which abstracts a mechanism for organizing information into a hierarchy, by holding group of files and child directories. That is, a directory structure organizes and provides information about all files in the system, mainly by connecting a file name with its unique identifier [1] (i.e., each directory entry consists of a file name and its identifier), being the latter used to locate the corresponding raw data of a file, as well as other file attributes specified by the metadata component.

Figure 4.19 shows a simplified view of directory structure, where each entry has a name and a unique ID, with the latter referring to the corresponding metadata node which points to the raw data of the corresponding file.

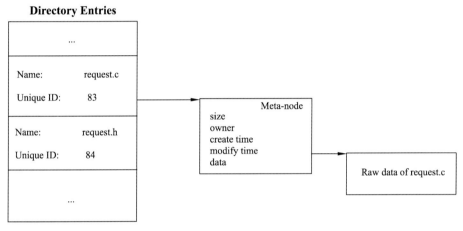

Figure 4.19 Simplified view of a directory structure, showing each entry content and how the metadata is accessed to locate the content of request.c

Usually, file systems leveraged by OSes like Windows and Linux follow the *everything-is-a-file* philosophy, in which not only user files but also devices are considered as files. In doing so, while defining properly a file, the above OSes offer the following basic I/O systems calls [2], which must be implemented by both file system and device drivers: creating, deleting, opening, closing, reading and writing a file. However, for a minimal file system paired [3] set of operations, other functionalities must be supported, such as:

1. *Initialization of a file system* normally creates an empty top-level directory (a.k.a., the root directory) on a given storage device;
2. *Mounting of a file system* consists of accessing a storage device, reading the superblock [4] and other file system metadata, and then preparing the system for accessing storage device;
3. *Unmounting of a mounted file system* involves flushing out to disk all in-memory state

[1] The unique identifier of a file is the so-called *i-node* number in Linux or *fileID* in Windows file systems. You can query this unique identifier using an OS service to locate the metadata component of a file, load that information and then access the corresponding raw data of a file.
[2] OSes normally offer a catchall *ioctl()* system call (i.e., a general-purpose mechanism) to manage any file system or device specificity, which does not fit into the standard I/O model.
[3] In practice, it is recommended that the usage of each individual operation be paired with its opposite one, e.g., (open, close), (create, delete), and (read, write), whenever possible.
[4] The superblock is located at the beginning of the storage device slice and it stores much of the information about the file system such as: size and status of the file system, file system name and name of storage device, size of the file system logical block, file system state, cylinder group size, date and time of the last update, number of data blocks in a cylinder group, path name of the last mount point, and so on.

associated with the corresponding storage device.

Examples of file systems supported by Windows and Linux are: (1) FAT, FAT32, FAT64 and NTFS in Windows and (2) ext2. Ext3, ext4, JFS and BTRFS in Linux.

The C file system leverages the concept of stream, an abstraction between programmer and a file, where a file is the actual device being used. The C file system is a buffered solution, implemented through functions like *fopen()*, *fclose()*, *scanf()*, *fscanf()*, *printf()*, *fprintf()*, *fgets()* and *fputs()*, contained in the standard C library (i.e., *stdio* library). So now comes the question: *how do streams and files interact?* A stream is a sequence of bytes that you can use to read from and write to a file, and it is created by calling the *fopen()* and closed only when the paired function, *fclose()*, is called with the corresponding pointer to the file descriptor returned by *fopen()*. The C File system delegates file management up to the execution environment, as it is layered on top of the minimalist set of I/O system calls that forms the OS I/O model. Buffering is used for temporarily storing data while data is being transferred to the OS kernel. It reduces the I/O processing overhead by minimizing the number of system calls. The concept of stream is generic as it is implementation- and machine-independent. Streams can be either text or binary and it may be created for reading (input streams), writing (output streams), or both (input/output streams). There are several kinds of streams, other than file streams, such as network, memory and pipe streams. A text stream is a sequence of characters while a binary stream is a sequence of bytes that has a one-to-one correspondence to the bytes in the storage device.

So far we have been using some functions of the C standard I/O library such as *scanf()* and *printf()* because whenever the OS launches your executable application for execution, it explicitly opens and associate three streams to the application, which allows you to communicate with your console (i.e., the screen and keyboard). The file descriptor 0 represents the input stream, *stdin*, which is connected to the input device like a keyboard, while the file descriptor 1 represents the output stream, *stdout*, which is connected to the output device like a display. A third error stream, represented by the file descriptor 2 is also opened and connected to the output device. Because of that, you can replace the statement "*scanf("%d",&selection);*" in Figure 4.3 with "*fscanf(stdin,"%d",&selection);*", as well as "*printf("Invalid argument%d\n",index);*" in Figure 4.16, with "*fprintf(stdout,"Invalid argument%d\n",index);*".

To exercise the theoretical introduction to file system let's imagine the configuration data is becoming very large and thus, unfriendly to be passed through the command line. The new revised problem statement will be: "*Compare a binary number, represented as a binary string, to the largest number found in a sequence of numbers, as required by a user who can specify not only his or her preferred conversion algorithm, but also many other execution settings. The result of the comparison will be registered for further analysis*".

Let's say the user wants to specify the number of requests to be processed along with the preferred binary converter callback. To accommodate any further increase of configuration data, let's use a configuration file approach to communicate with the executable program during setup-time. Since different users may want different configuration settings, let's also specify the configuration file through command-line. Although one may choose between using C standard I/O or straight OS I/O system calls, let's choose the former to better consolidate the introduced knowledge while gaining functional skill in using the C file system.

The function that previously validates the configuration setting, inputted via command-line arguments, was accordingly refactored to parse the configuration file line-by-line and then retrieve all configuration parameters (see *check4ValidConf()* in Figure 4.20). The configuration file has a very simple format, one parameter per line, in the following order:

1. The callback index for the first request;
2. The callback index for the second request;
3. The number of requests to be processed;
4. The result file name where the program results will be directed to.

Looking at the first two red-marked blocks on *Watch 2* window of the Figure 4.20, you will see the raw data of *conf2.txt* as a set of four strings separated by '\n', as shown on the Table 4.1.

Figure 4.20 Execution state of programming algorithm for the refactored check4ValidConf(), showing the content of the inputted configuration file through an opened stream represented by the file pointer variable, pfDescriptor

Table 4.1   Configuration Settings: conf2.txt

0	1st callback index
1	2nd callback index
2	No. of request to be processed
Result2.txt	Result file name

The file pointer variable, *pfDescriptor*, maps to a file descriptor which is an implementation of the generic meta-node discussed above and represented in Figure 4.19. That is, the C standard I/O functions operate on file pointers instead of on file descriptors. Basically, *fopen()* associates the file *conf2.txt* with an input stream and then it initializes *pfDescriptor* with corresponding *conf2.txt* metadata. To transfer the data from the storage device to memory, *fscanf()* was called with *pfDescriptor* as its first argument. The *Watch 2 window* in Figure 4.21 illustrates the behavior of *fclose()* which completely cleans the previously allocated file pointer, *pfDescriptor*, while returning its associated file descriptor back to the OS.

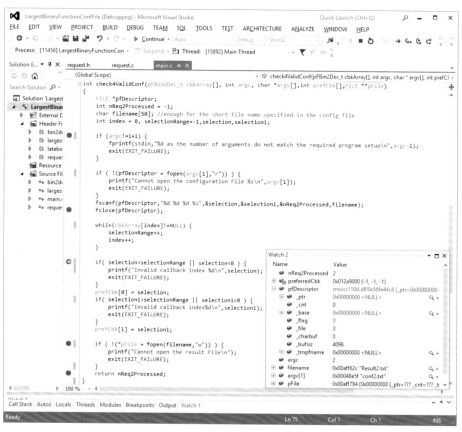

Figure 4.21   Execution state of programming algorithm for the refactored check4ValidConf(), showing the content of the file pointer variable, pfDescriptor, after calling fclose() to return the handle to the OS

Looking at the rightmost formal parameter of *check4ValidConf()*, marked with the uppermost blue block on Figure 4.21, you may ask why use a pointer to file pointer instead of simply a file pointer. Try it and you will see the program crashing because pointers are by default passed by value. Therefore, to pass a pointer by reference one needs to pass a pointer to the target pointer (i.e., *FILE **pFile* instead of *FILE *pFile*).

Looking at Figure 4.21 and Figure 4.22, it is visible the recommended practice of the paired usage of *fopen()* and *fclose()*, as shown by the pairs of blue- and green-marked statement blocks.

The red-marked block on the *Watch 2 window* of Figure 4.22 shows the in-memory update of *Result2.txt* during the execution at the body of the "*case 2:*" switch statement, i.e., updating the output stream with the result of the first processed request. Similarly, Figure 4.23 illustrates the final output data transfer corresponding to the processing result of the second request. Basically, *Result2.txt* contains one line of text for each processed request.

Notice that everything presented so far during this debugging process, should be started by specifying *conf2.txt* as the command-line argument. As previously done, you only need to enter the name of the chosen configuration file at *Command-line arguments* field, as shown in Figure 4.24.

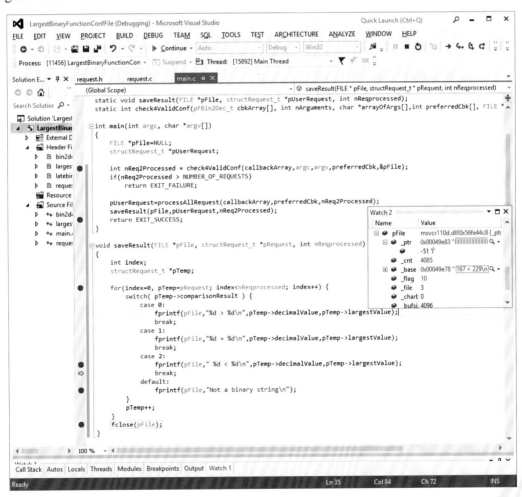

Figure 4.22 Execution state of programming algorithms for the refactored main() and saveResult(), showing partially data transfer over an output stream to save the result of the first processed request

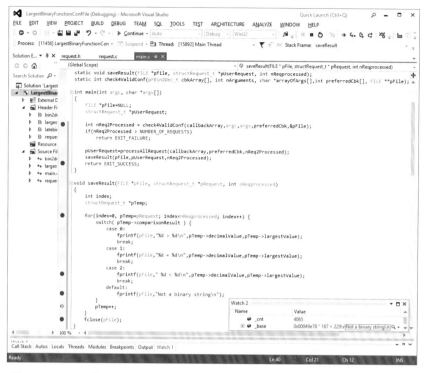

Figure 4.23   Execution state of programming algorithms for the refactored main() and saveResult(), showing the final data transfer over an output stream to append the second request processing result

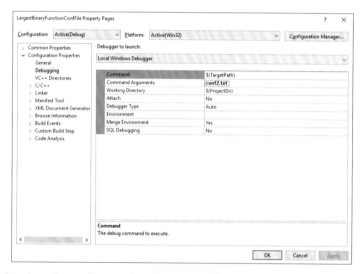

Figure 4.24   Inserting conf2.txt as the only command-line argument with the program settings

Figure 4.25 presents the refactored programming algorithm for *processAllRequest()*. Algorithmically, it is virtually identical to the previous version. Only its signature was extended to incorporate another formal parameter, with the corresponding passed argument retrieved from the inputted configuration file. According to Table 4.1, such argument will be given by the third entry of the configuration file, which dictates the number of requests to be processed, starting from the first one on the structure array variable, *allRequests*. The lowermost blue-marked block shows how the loop is changed to accommodate this new parameter, instead of using the constant total number of requests, *NUMBER_OF_REQUESTS*. In Figure 4.26 the blue-, green- and red-marked blocks show the configuration files, the passing of the command-line argument and the result files after each executable program run, respectively.

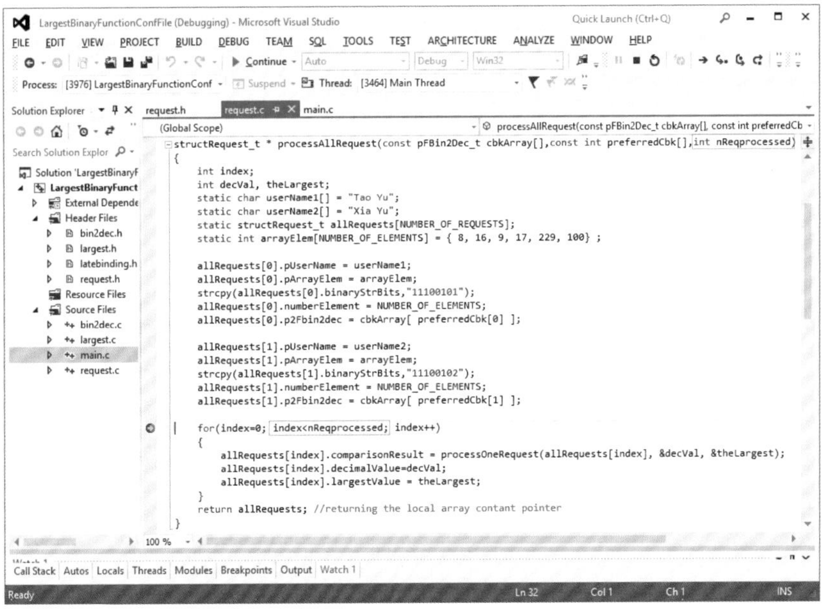

Figure 4.25   The programming algorithm for the refactored processAllRequest(), showing the new signature extended with the formal parameter nReqprocessed

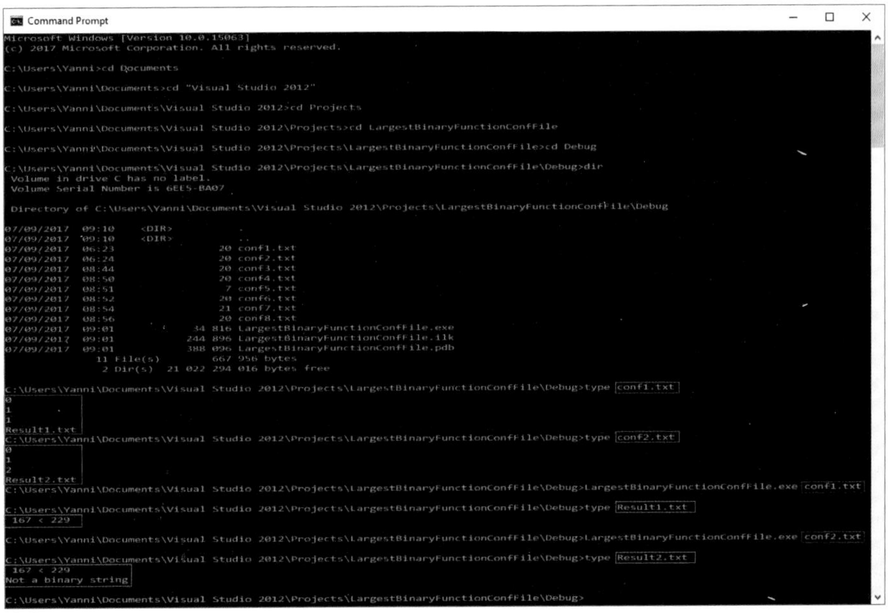

Figure 4.26   Executing the executable program from DOS command line by passing directly the chosen configuration file

Finally, to explore and discover how the C file system delegates file management up to the OS, remove first all breakpoints and insert only one on the blue-marked statement in Figure 4.20 and then start debugging following the steps below:

1. Press F11 to step into the callee body;
2. Press successively F10 until you reach another call site;
3. Repeat the above steps until you get back to the statement after the initial *fopen()* call site.

In doing so, you will see how the interface for Windows OS services (i.e., the specialized Windows File Management API) are used through several low-level helper functions such as _tfopen(), _tfsopen(), _openfile(), _topenfile(), _tsopen_s() and _tsopen_helper(), instead of directly call the corresponding *open()* system call. Normally, the C file system functions in the standard C library provide additional capabilities (e.g., buffering mechanism) not available directly through system calls. For example, reading a file using library functions, usually read much more than the requested amount of data, with the extra byte buffered to satisfy further read requests by the application, and so, minimizing the number of system calls.

4.4 Dynamic Memory Management in C and Linked List

Previously in Chapter 2, you learned that C leverages three types of lifetime (i.e., static, automatic and dynamic), leading the memory of applications to be statically, automatically, or dynamically managed. Furthermore, a generic C program memory layout split into several memory areas, which is almost completely dictated by compiler-based tools and OS program loader, was presented in Figure 2.31. So far, you have learned and practiced the predictable compiled-based allocation of variables on data and stack segments (i.e., automatic- and static-allocated variable), as dictated by their lifetime. Therefore, you may raise the following question: *is the compiled-based variable allocation adequate for all situations?* Recall automatic variable allocation on stack segments does not persist across multiple functions, while static variable allocation on data segments persists for the program lifetime, whether it is needed or not. For instance, what if the users of the last discussed and implemented program demand for a variable number of conversion-comparison requests to be processed, altogether with a variable number of callback binary converters to be chosen from. In this case, you can declare a structure array of requests, *allRequests[]*, and a function pointer array, *callbackArray[]*, both large enough to accommodate any number of request and callbacks, respectively. Maybe it is not possible to anticipate how large these arrays need to be or if possible, it will be a huge waste of memory for most execution scenarios. Since the sizes of both arrays depends on information that is only available at runtime, then you need to postpone the decision on both sizes, until the program runs. This is the principle of dynamic memory management which allocates memory in size that may be unknown until you actually need it at runtime, and release it again as soon as possible. That is, dynamic memory allocation fits really well when the amount of needed memory or how long you might need it, varies and it is known only after the program is run.

The dynamic memory allocation uses the heap segment, whose size is initially set on application startup, but it can be expanded at run-time by using OS services. Furthermore, through OS services, it uses pointer variables to request memory which size and lifetime are completely unpredictable. Before presenting specific details about dynamic memory allocation implementation and usage, Figure 4.27 depicts an overview of the memory management model, as deployed on modern computer systems, and how dynamic allocation fits on it. This model bears some analogy to that of wholesale business [1], with OS memory managers, heap managers and applications representing wholesalers, retailers and end users (i.e., standard memory consumers), respectively. The processor-based hardware can be seen as a manufacturer as it provides the physical memory.

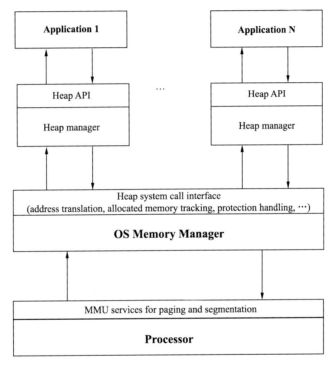

Figure 4.27. A generic memory management model

Based on this model, memory management really occurs at two-level with a single operating system memory manager and different heap managers specific for each running application:

1. At the OS-level, it is carried out by the memory manager of the corresponding OS and it consists of keeping track of what application owns what memory block, allocating more memory to an application when requested, and protecting each application set of memory blocks from other applications. The operating system manager allocates rather large blocks of memory, usually at page granularity (i.e., few Kbytes blocks), to individual heap managers. For performance reasons, the OS memory manager and the processor work closely together. The processor provides paging and segmentation services for accessing, protecting, and emulating memory, while the memory manager decides how to use these

---

[1] In a wholesale business model, the manufacturer produce products while a *wholesaler* is an intermediary entity in the distribution channel that buys in bulk and sells to retailers rather than to consumers.

processor's services to perform repetitive bookkeeping down on the hardware. Memory management at this level faces challenges such as (1) executing programs of arbitrary size, (2) having several programs in memory at the same time and (3) freeing programmers from memory management tasks as much as possible. In doing so, it carries out the following tasks: translating addresses, allocating memory to load applications, allocating memory for applications, tracking application memory allocations, tracking physical memory, and handling protection violations, just to name a few.

2. *At the application-level* is performed by the heap manager that come with each application and it consists of keeping track of free smaller blocks of memory (i.e., at word granularity) or reclaiming those blocks that the application no longer needs. That is, for the heap manager, dynamic memory allocation includes requesting large memory blocks from the OS memory manager if needed, deciding which block to allocate, splitting it into smaller blocks as required for the application, and removing it from the list of free blocks. The usage and implementation of these dynamic allocation functions are language-dependent and application programs use them to manage and clean their own heap segments.

Unlike Java and C# languages that rely on an automatic heap management supported by a virtual machine, C language provides almost no abstraction over memory. Therefore, the heap management is programmatically done under programmer's responsibility, by calling *malloc()*, *calloc()*, *realloc()*, and *free()* whose prototypes are located in *stdlib.h*. Similarly to the C file system explained above, these C standard functions provide OS- and platform-independent interfaces that use the heap manager through its specialized interfaces and are then forwarded to appropriate heap system calls.

To exercise the theoretical introduction to dynamic memory allocation, let's say users demand for a variable number of conversion-comparison requests to be processed, with some of them generated while the program is running.

The new revised problem statement will be: "*Compare a set of binary numbers, represented as a binary string, to the largest number found in a sequence of numbers, as required by a user who can specify not only his or her preferred conversion algorithm, but also many other execution settings. The set of requests can grow during program execution, if desired by the users. The result of the comparisons will be registered for further analysis*".

To accommodate the demand for variable number of requests during runtime, the following tasks will be performed:

1. The previously defined structure type, *structRequest_t*, will be slightly extended with a field pointing to itself, enabling the implementation of linked list;
2. The *main()* will be refactored to allow further request processing during runtime;
3. A new module will be provided for linked list management, offering creation, insertion and deletion functions.

The file header in Figure 4.28 shows the extended structure type with the pointer variable field, *pNext*, which serves as a node connector (see the uppermost blue-marked block). A linked list depicted in Figure 4.29, is a linear data structure consisting of a collection of data elements or nodes, in which the linear order is given by connectors pointing to successor or predecessor nodes. In C language the connector, *pNext*, is represented by a pointer variable of the same type as a node (e.g., *pNext*), being the last connector always pointing to NULL.

Figure 4.28   The proposed linked list node defined as a structure data type and the operations for its management

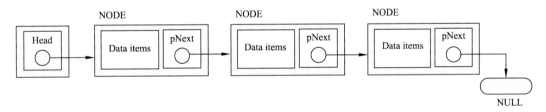

Figure 4.29   Graphical representation of a linked list, with nodes and connectors

The lowest blue-marked block in Figure 4.28 shows functions prototypes representing the basic operations performed on the linked list. The linked list is represented by the head node, *pUserRequest*, based on the structure type *structRequest_t*, as shown above by the uppermost blue-marked statement block in Figure 4.30. The new *main()* was refactored into the next three lowest blue-marked statements blocks. From up to down, these three statements blocks process the initial users' requests according to the setup configuration of the application, update the linked list with new request before processing them all, and save results to file before cleaning the heap by returning back to the OS all previously allocated memory, respectively. On the *Watch 4* windows you will see that automatic or local pointer variables *pUserRequest* and *p* are not initialized, according to their assigned addresses.

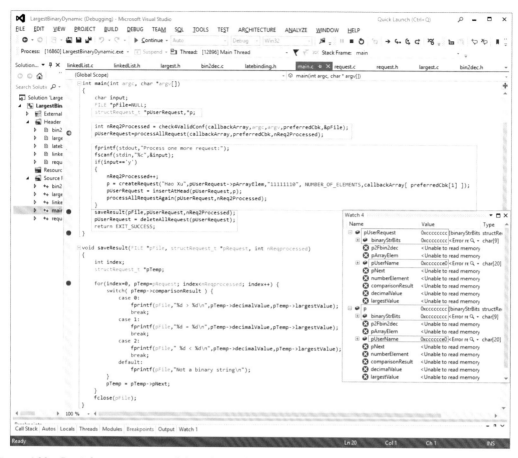

Figure 4.30   Partial execution state of the refactored main() and saveResult(), showing initial state of the linked list whose head node is pointed by the structure pointer variable pUserRequest

The only change made to *saveResult()* is the use of the node connector while scanning the linked list (see the lowest blue-marked statement in Figure 4.30). Figure 4.31 and Figure 4.32 present the linked list management functions for single node creation and deletion (i.e., *createRequest()* and *deleteRequest()*), full list deletion, *deleteAllRequest()*, as well as the insertion of a new node on the list at the head, *insertAtHead()* and at the tail, *insertAtTail()*.

The uppermost blue-marked statements block on the body of *createRequest()* in Figure 4.31 illustrates the usage of *malloc()*. Firstly, it is used to allocate a block of memory on the heap by taking as a single argument, the amount in bytes of the memory to be allocated. Secondly, since *malloc()* returns a pointer to a void type (i.e., void *), it can be assigned to any type of C pointer, as C automatically promotes pointers to void to any other pointer type on an assignment. We typecast it to *"structRequest_t*"* only for later compatibility with and porting to C++, as the latter does not perform automatic *void* pointer promotion. Just remove the typecast and you will see that everything will be fine. Thirdly, if you insert a breakpoint at the statement immediately after the *malloc()* call site, i.e., at *"if(!pReq)"* statement, while observing *pReq* on an *Watch* window, you will see that *malloc()* does not initialize the allocated memory. Please try it by yourself. Therefore, the in-the-middle blue-marked block of statements in Figure 4.31 shows how data items of the created request node are set before being processed. For instance, the connector structure pointer field, *pNext*, should always be set to NULL. Still on the in-the-middle statements block you will see that any node field not statically allocated, e.g., *pReq->pArrayElem*, should also be dynamically allocated and initialized, if needed. Fourthly, uppermost and lowest blue-marked statements blocks show that all calls to *malloc()* should be checked and in case of a failure, the program execution must be ended. Although the possibility of failure in allocating memory is small, a call to *malloc()* fails if: (1) there is not enough memory left for allocation, (2) the size requested exceeds the limit allowed for allocation or (3) the heap memory manager has somehow been corrupted.

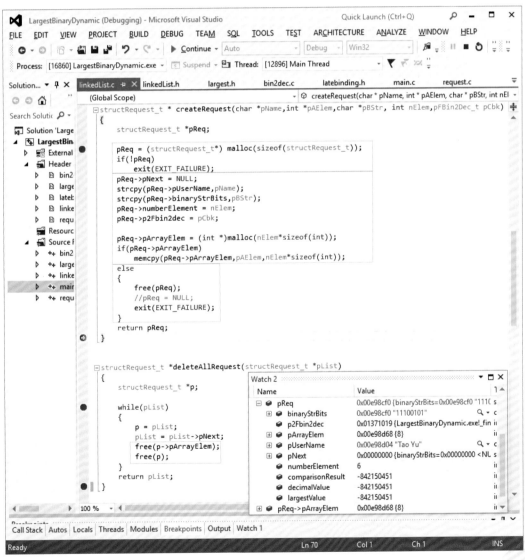

Figure 4.31   Partial execution state of createRequest(), showing the dynamic allocation of a new node and updating of its data items, as well the programming algorithm for deleteAllRequest()

The two lowest blue-marked statements blocks in Figure 4.31, as well as *deleteRequest()* in Figure 4.32 illustrate the usage of *free()* to deallocate a block of memory that is no longer needed. The passed argument to *free()* must always be a pointer variable previously allocated by calling *malloc()*, *calloc()*, or *realloc()*. Otherwise, it can cause bugs, including the corruption of the heap manager. Another problem will happen if the pointer to a memory block has meanwhile been deallocated. Thus, the pointer variable, *pRequest*, was explicitly nullified in the blue-marked block in Figure 4.32 to designate it as an invalid pointer. Notice also the order in which the two *free()* were called. That is, in the reverse order of *malloc()* calls in *createRequest()* as all dynamically allocated items of a node should be freed before the node itself. The programming algorithm for *insertAtTail()* in Figure 4.32 illustrates how the linked list, *pList*, which is passed by reference, is scanned until the nulled connector is found and then pointing it to the passed node argument, *pElement*. Otherwise the linked list is empty and thus, it will point directly to the fresh inserted node. Looking at the *Watch 2* window, you will see the connector of the first and previously inserted node is pointing to NULL, as shown by the red-marked debugging block. The programming algorithm for *insertAtHead()* in Figure 4.32 illustrates how a new node, *pElement*, is inserted into the linked list, *pList*, which is passed by value. In case of a not empty linked list, the node passed as argument becomes the head of the linked list, but first its nulled connector will be pointed to the previous head of the linked list.

Both *processAllRequest()* and *processAllRequestAgain()* in Figure 4.33 process the demanded number of requests as specified during setup-time, and they defer only because the former starts by populating the initial linked list pointed by the structure pointer variable, *pReqs*. The latter will be called later when a new node is created and inserted in the linked list, as shown below by blue-marked statements block in Figure 4.34. The uppermost blue-marked block in the Figure 4.33 illustrates the use of *createRequest()* and *insertAtTail()* to initially populate the linked list pointed by the structure pointer variable *pReqs*. The two lowest blue-marked statement blocks show how each node's connecter, *pNext*, is used inside the two loop bodies to pass from one node to another until the nulled connector is reached. The *Watch 2* windows shows the data items of a node after being processed.

Looking closely at the *Watch 3* window and more specifically to the assigned addresses of the six variables, you will see that the two global variables, *preferredCbk* and *callbackArray* are on the data segment, *pUserRequest* and *p* which are dynamically allocated are on the heap, and finally the two local variables *input* and *nReq2Processed* are on the stack, exactly according to the functional organization of a C program when loaded into memory, as previously discussed in Chapter 2 and shown in Figure 2.31.

4 HANDS-ON-POINTERS: ADVANCED FEATURES • 177

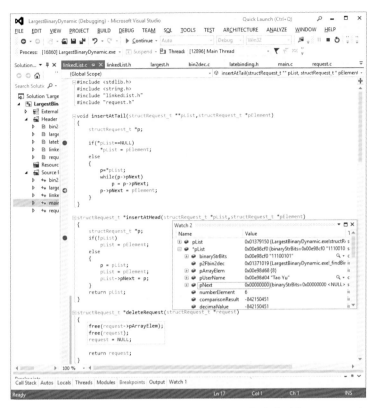

Figure 4.32   Partial execution state of insertAtTail(), showing how the linked list is scanned to insert the second node to end the linear sequence, with only one already inserted node

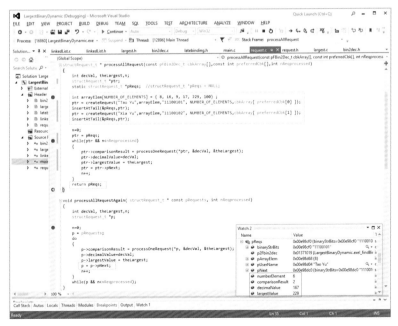

Figure 4.33   Execution state of processAllRequest(), showing how the linked list is populated with two initial nodes by calling createRequest() and insertAtTail() and then scanned using node connectors for processing individually each node

Figure 4.34   Partial execution state of main(), showing assigned addresses to statically, dynamically and automatically allocated variables in data, heap and stack segments, respectively

If you look at the *Output* window after compiling the program presented above and some other previous versions, you will realized that the MS Visual Studio C compiler claims for an implementation that is conformed to C11 standard, as shown by the Figure 4.35. The MS Visual studio C/C++ environment used throughout this book widely implements the C11 standard Annex K functions.

The Annex K library introduces new functions that leverage more secure programming through a kind of bounds checking interface (e.g., some of those functions check arrays' bounds in a data copy operation). Compared to traditional standard I/O functions, the name of alternatives functions offered in the Annex K library come with a postfix _s, such as *fscanf_s()*, *fopen_s()* and *strcpy_s()*, as illustrated by the paired red-marked blocks in Figure 4.35. C11 standardizes the semantics of these _s functions which promote security benefits through the following four patterns[1] of:

1. Least privilege;
2. Minimize TOCTOU[1] vulnerability;
3. Reduce the return value variability using *errno_t* returned values;
4. Use runtime constraint handlers for logic errors.

---

[1] For more details see "*The New C Standard Explored*" by Tom Plum, Dr. Dobb's, May 08, 2012.
[1] TOCTOU ≡ Time-Of-Check versus Time-Of-Use

4 HANDS-ON-POINTERS: ADVANCED FEATURES

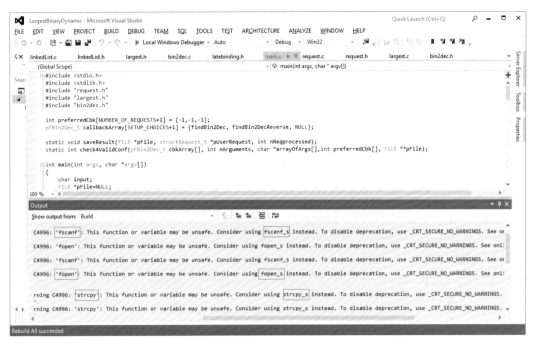

Figure 4.35   Compiler warnings regarding non-C11 conforming user-written code

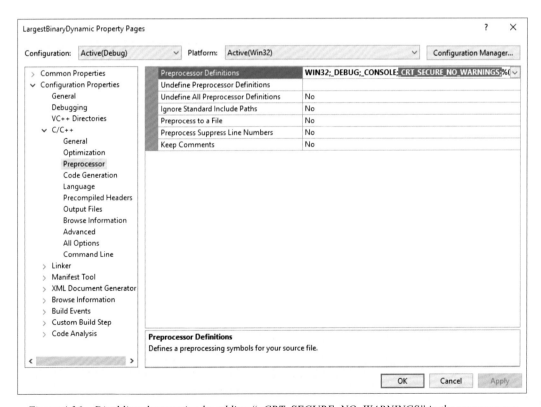

Figure 4.36   Disabling deprecation by adding "_CRT_SECURE_NO_WARNINGS" in the preprocessor definitions

You can eliminate secure warnings (i.e., disable deprecation) from the project by simply defining `_CRT_SECURE_NO_WARNINGS` through the project property page (see Figure 4.36). Alternatively, you can opt for secure *programming* and then replace traditional standard I/O functions by their equivalent *_s* pairs.

As homework you should change the above program to accommodate a variable number of binary converters to be chosen from, with the callback table updated while the program is running. Additionally, you should study and practice the usage of *calloc()* and *realloc()*, as well as some *_s* functions.

This chapter's main focus was on some advanced pointer features and it will end now with the following recommendations:

Recommendation 23: Always remember *main()* as representing a complete C program and when the OS executes a generated executable file, it actually executes *main()*.

Recommendation 24: Always pair *malloc()* with *free()* throughout the program to avoid memory leaks which occur when memory is allocated with *malloc()* and never delete after being used.

Recommendation 25: Always decide carefully about the usage of dynamic memory as its greater flexibility can come with a cost, mainly for programs where performance and determinism are main concerns. Using static memory improves program' determinism while allowing applications to execute faster, since it has not to perform any extraneous bookkeeping at runtime. That is, the higher the number of allocations and deallocations, the poorer the program's performance and determinism.

Recommendation 26: Since the contents of allocated memory returned by *malloc()* are arbitrary and hence meaningless, always initialize it accordingly before any further use.

Finally, to evaluate your debugging skills, the above program deliberately came with some silent bugs and we shall challenge you to pinpoint and fix them, while identifying which above recommendations were not fulfilled.

References

[11] SILBERSCHATZ ABRAHAM, GALVIN PETER B. Operating System Concepts. Hoboken: Wiley, 2012.

[12] BLUNDEN BILL, Memory Management, Algorithm and implementation in C++. Plano: Wordware publishing, 2003.

[13] FRANEK FRANTISEK. Memory as a Programming Concept in C and C++. Cambridge: Cambridge University Press, 2004.

Several other information sources were also used, mainly from internet, as well as those previously referenced in previous chapters. Thus, the credits also go to all them.

5 FROM C TO C++

Learning objectives	1. Understanding Object-Oriented Programming (OOP) paradigm. 2. Understanding of C++ key concepts. 3. Understanding key advantage while migrating from C to C++. 4. Understanding linkage to C code. 5. Practicing refactoring from C to C++ by example. 6. Motivating students for mastering C++ by themselves.
Theoretical contents	1. A brief overview of C++. 2. Usage of some C++ key concepts. 3. Refactoring of small-sized C program to C++.

Strategies and activities	1. Introduce and explain during refactoring from C to C++ simple, immediate, hands-on experience with key features of C++ such as:
	  a. Stream I/O.
	  b. Function overloading.
	  c. Reference parameters.
	  d. Classes.
	  e. Constructors.
	  f. Destructors.
	  g. Access modifiers.
	  h. Namespace.
	  i. Exception handling.
	2. Illustrate how C++ implements data hiding and encapsulation mechanisms as some of Object-Oriented Programming concepts.
	3. Refactoring a previously implemented small-sized C program to C++, while demonstrating how C++ can deal with the increase in complexity of programs:
	  a. Exercising on creating and destroying objects.
	  b. Exercising on reusing implementation through composition.
	  c. Exercising on reusing interface through inheritance.
	  d. Practicing with polymorphism and static member function.
	  e. Practicing with namespace, friend mechanism and exception handling.

C++ is a hybrid object-oriented programming language built upon the foundation and syntax of C language, to remove some marketplace pressure faced by the latter, mainly due to built-in limitations regarding to program size and complexity. Therefore, C++ not only preserves the C efficiency, but enables better and easier management, as well as evolution of todays' programs increase in size and complexity. Such evolutionary boldness is supported by its object-orientation and has been also progressively enhanced through generic/template and functional programming paradigms, as well as through other concepts such as namespace and exception handling. Furthermore, comparatively to C language which is type-safe only in limited contexts, C++ presents a stronger type-safe model. Regarding strictly to object-orientation, C++ leverages the following features of pure OOP: (1) everything is an object, (2) every object is an instance of an abstract data type (ADT), (3) every object has its own memory made up of other objects (4) all objects of the same type receive the same messages and (5) a program is a set of communicating objects. An ADT is implemented and expressed in C++ using the class keyword, which packs data and operations allowed on data together by concept. The kernel language developed from C is classically used as a system-implementation language, while the class-based additions to the language support the full range of OOP requirements.

5.1 A Brief Overview of C++ Main Features

In general terms, C++ is a successor and superset of C, extended through several features rest on the ability to create new ADTs (i.e., classes), and thus, promoting superior code organization. Technically, C++ supports object-orientation through ADT which is an encapsulation mechanism, along with inheritance and polymorphism (i.e., runtime type binding). Contrary to C that organizes programs around code (i.e., a program is defined by its functions which can operate on any of its used data), C++ as an object-oriented programming language organizes program around data, by first defining data and then the methods that are applied on that data.

Generically, ADTs in C++ are expressed by the class concept, which consists of data types and a set of methods, as well as an implementation hidden from the programmer who uses the data types. It primarily accommodates the increasing complexity of programs by distinguishing and isolating an object's internal state and behavior from its external state and behavior (i.e., encapsulation with data hiding), while creating new user-defined types that fit the needs of the problem to be solved (i.e., type extensibility). To hide internal object details and protect object integrity, C++ offers three access specifiers, given by explicit keywords of *public*, *private* and *protected*. Furthermore, the implementation hidden enables the change of internal state and behavior of a class without affecting the users of a class. The set of methods that are not part of the hidden implementation are denominated as interface and it establishes the requests the users of class can make for a given object of that class. To fully support the creation and usage of ADTs, the concepts of inheritance and polymorphism are vital.

The concept of inheritance is a code sharing and reuse mechanism which means that one class gets traits from another class. That is, a generic class can be defined with common traits to a set of related items, and then used to create derived and more specific classes, each adding to the generic class only those features that makes it unique within that class family. It is designed into software to maximize reuse, as well as to allow a natural modeling of the problem domain, by developing a hierarchy of related ADTs that shares code and a common interface. According to standard C++ terminology, a class that is inherited is referred to as a base class and the class that does the inheriting is called the derived class. Further, a derived class can be used as a base class for another derived class, developing a hierarchy of related ADTs through multiple inheritance.

The concept of polymorphism depends on inheritance since the latter enables the creation of a family of ADTs, while objects in that family are manipulated through their common base class interface. Poly means many, and morph means form and so, polymorphism is usually characterized by the phrase *"one interface, multiple methods"*. It always comes into two types: ad hoc polymorphism (e.g., coercion and overloading) and pure polymorphism (e.g., inclusion and parametric polymorphism). Coercion, overloading, inclusion and parametric polymorphism (e.g., template) enable a function/operator to work on several types by converting their values to the expected type, function calls based on their signatures/prototypes, ADTs to be derived from another ADT with methods available for the base ADT working on derived ADTs, and an ADT to be unspecified and later instantiated, respectively. According to the inclusion mechanism, at each call site the C++ runtime system automatically determines the derived ADT and consequently the right method to be invoked. Since C++ is a compiled language, it supports both runtime and compile-time polymorphism, which means the programmer only needs to remember and use the family interface, while the compiler statically and dynamically decides for the method to be invoked. Summing up, polymorphism helps reduce complexity by allowing the same interface to be used to access a family class of methods, while improving code organization and readability as well as the creation of extensible programs when new features are desired.

However, other features are needed to complement C++ OOP requirements, such as creating and destroying objects, as well as exception handling and namespaces:

1. To create and destroy objects, classes provide two member functions, the constructor and the destructor so that a user of a class can use objects like native C types. The constructor has the same name as a class and it mainly initializes data members and allocates storage from heap, if needed, using the new operator. The destructor shares also the class name but prefixed by the tilde character, ~. Its main purpose is finalizing or destroying objects of the class type, using the delete operator if the object or its data members are created on the heap.
2. Exceptions are unexpected runtime error conditions automatically managed by an

exception handling subsystem, instead of coding it manually in the program. Such runtime errors which usually terminate the user program with a system-provided error message, will automatically trigger an error-handling routine, especially design for a particular type of error. The error-handling routine does not interfere with normally-executing program code, as it uses a separate execution path. Its main purpose is fixing issues and restoring the program execution. C++ exception handling subsystem is programmatically supported by the try block statement built upon try, catch, and throw keywords. Statements to be monitored for exceptions are contained in a try block, occurrence of errors within the try block are thrown, and errors are caught using catch and then fixed (i.e., processed).

3. The *namespace* statement helps organizing complex and large programs, by creating a declarative region to localize the names of identifiers. It avoids name collisions, since elements declared in one namespace are separate from elements declared in another. That is, it defines a scope to differentiate similar functions, classes, variables and so on, with the same name available in multiples and different libraries, as well as in program files coded by different persons. Most importantly, C++ namespace scope eliminates the tedious, time-wasting and expensive activities needed by developers to avoid accidental use of the same names in situations where they can conflict. Besides the *namespace* statement, C++ offers the *using* directive to bring a namespace scope into global visibility, and thus, enabling better control over identifiers in terms of their creation, visibility, placement of storage, and linkage.

5.2 Usage of Some C++ Key Concepts

In this paragraph several C++ key concepts will be practiced while the last C program developed in Chapter 4 is refactored to C++. The separation in declaration and implementation files will be persevered as before, to easily promote reusable classes. However, there is a clear difference from C, as implementation files with the definition of member functions are stored as *.cpp* files while declarations are still stored in header files with *.h* extension. Preferentially, both kinds of files should share the same name as the class name.

Instead of directly mapping of previous *C.h* to classes declaration files and *.c* files to *.cpp* classes implementations, we approach the refactoring process by first identifying which elements of the problem space can be translated as ADTs and then decide on an appropriate set of classes that better fit the binary comparison-conversion problem. Figure 5.1 to Figure 5.4 present class declarations of the four identified ADTs.

The first look at the file declaration for the abstract class *ARequest* in Figure 5.1 will take us to the *const* modifier, as shown by the uppermost blue-marked block. It replaces the previous preprocessor command #*define* to make both initialized variables, *NAME_LENGTH* and *NUMBER_OF_BITS* non-modifiable, i.e., making them both symbolic constants. The *ARequest* class is an ADT or user-defined type in which the hidden implementation is assigned an access control qualified by the keyword *protected* to avoid data members from being manipulated by the users of the class. The privacy specification of *protected* is assigned instead of *private* in order to give the inheriting class, *CRequest* as illustrated in Figure 5.2, access to these protected data members. In other words, protected members are private members with special rules when they are used by a derived class.

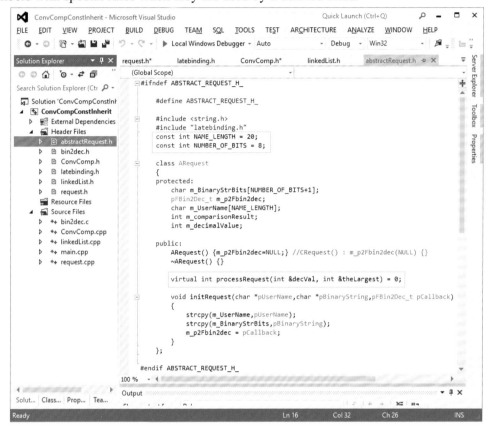

Figure 5.1   Declaration of an abstract base class, ARequest, for a possible family of conversion requests derived classes

To be usable, objects of a class must presented an interface which the user of the class can access to command it to perform its operations. For the class *ARequest*, the interface is given by the constructor, the destructor and member functions, i.e., *processRequest()* and *initRequest()*, under the privacy specification of *public*.

The public inheritance of *CRequest* from *ARequest*, illustrated by the uppermost blue-marked statement block, creates a type hierarchy in which the interface of the base class, *ARequest*, will be also part of the interface of the inheriting class, *CRequest*. Looking back at Figure 5.1 and the lowest blue-marked statement block, you will see that the *ARequest* declares a pure virtual member function, *processRequest()*, under pure polymorphism through the inclusion mechanism, which is later redefined in the child class, *CRequest* (see the lowest-blue-marked statement in Figure 5.2). In doing so, the implementation of the member function *processRequest()* is undefined in *ARequest* and deferred until the implementation of *CRequest*. Therefore, *ARequest* becomes an abstract class (i.e., a class with at least a pure virtual function) which cannot be used to declare objects, although it can be used to declare pointers or references to access objects of its inheriting classes, such as instances of *CRequest*.

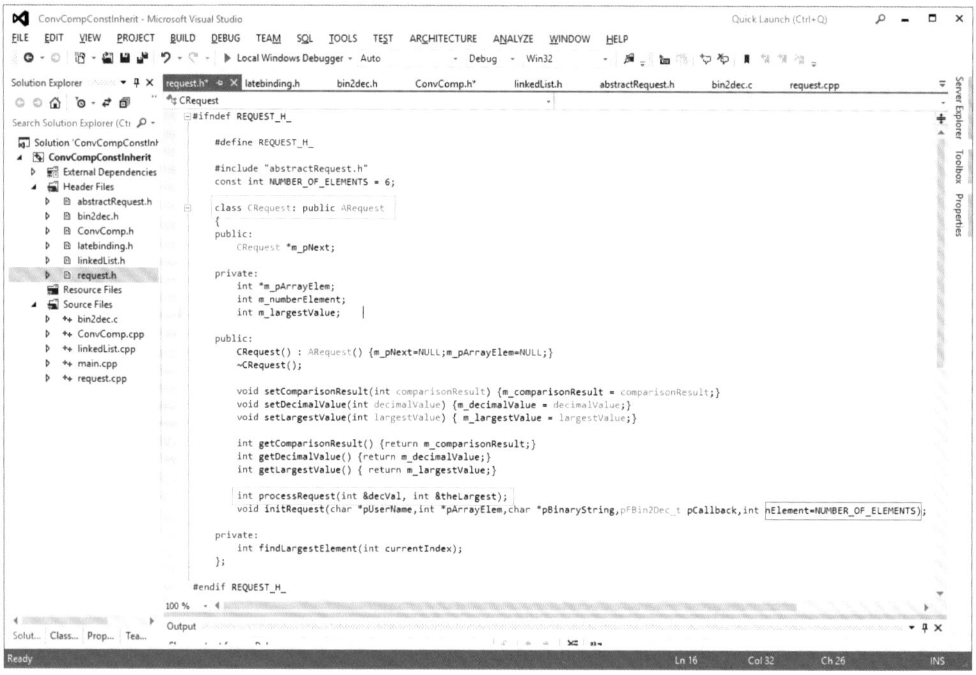

Figure 5.2   Declaration of an inheriting or derived class, CRequest, from the base class ARequest, promoting reuse of the interface

Contrary to the *CRequest* class which promotes the reuse of interface through inheritance (i.e., under a "*is-a*" relationship), the class *CLinkedListRequest* illustrates the reuse of the implementation of class *CRequest*, by creating private member objects *m_pListHead* and *m_pListTail*. That is, *CLinkedListRequest* is composed by two objects of previous class *CRequest*, under a "*has-a*" relationship. The two pointers to instances or objects of the class *CRequest*, as shown by the uppermost blue-marked statements block in Figure 5.3, are private data members because classes have a default privacy specification of *private* (i.e., if not explicitly qualified with a different access specifier).

Contrary to the way constructors of *ARequest* and *CRequest* initialized in their bodies some of their data members, the lowest blue-marked statements block in Figure 5.3 illustrates an alternative approach for the initialization of data members, using the special syntax of constructor initializer. Constructor initializer for class members are specified by a colon and a comma-separated list following the constructor parameter list and preceding the code body. Basically, it is a data member identifier followed by a parenthesized expression.

In Figure 5.3 the header *<fstream>*, which defines classes such as *ifstream*, *ofstream*, was included to perform file I/O operations, as required by the member function *saveResult*(). Furthermore, since all C++ I/O header files are wrapped in the namespace *std*, the *using* directive will be needed to bring *std* namespace into visibility. In so doing, it will allow access to symbols in the standard C++ library without prefixing them individually with "*std::*" (i.e., no need for scope-resolved names).

Figure 5.3   Declaration of the class CLinkedListRequest, promoting the reuse of an implementation through composition

In Figure 5.4 the *extern "C"* is used to enable the linkage to code in C, by commanding the compiler to generate calls to *findBin2Dec()* and *findBin2DecReverse()* using C calling convention. By default functions are linked as C++ functions and the linkage specification forces the two above functions to be linked in the final program as C functions, instead. The need for the linkage specification is justified due to the mixed compilation mode, as "*bin2dec.c*" contains C code, while remaining implementation files contain C++ code. Pointers to C functions are used to define an array of pointer to functions, as marked by the uppermost blue statement block in Figure 5.5. In doing so, any minor incompatibilities between C and C++ will be correctly fixed, mainly the one due to name mangling. C compilers do not mangle names (i.e., only one function of a given name in a scope can have C linkage) as C++ compilers do to support function overloading. Thus, the linkage to C should be used to prevent name mangling of C functions such as *findBin2Dec()* and *findBin2DecReverse()*. Otherwise, calls to both functions from C++ code will reference a mangled form of their names, ending with the linker complaining about external unresolved symbols.

The lowest blue-marked block of statements of Figure 5.5 presents you with alternative ways to create an object of *CLinkedListRequest* class using the *new* operator, which directly replaces the C standard library function *malloc()*. This is possible because in *CLinkedListRequest* class the constructor is overload through a default constructor (i.e., the one receiving no parameter) and another one receiving two pointers to objects of *CRequest*, as shown in Figure 5.3. A possible reason for the overloading of a constructor is to enable objects' creations according to particular scenarios. All data members of *CConvComparison* class have by default the privacy specification of private.

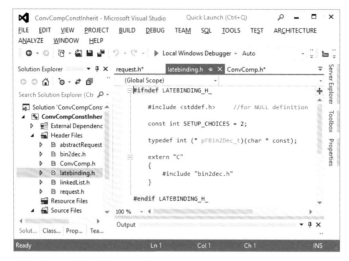

Figure 5.4 Linkage to C code to avoid the linker complaining regarding unresolved external symbols included in "bin2dec.h"

Figure 5.5 Declaration of the class CConvComparison exercising C++ I/O as well as linkage to C and implementation reuse

Member functions like the accessors *getSaveFile()*, *getReqprocessed()* and *getUserRequests()* are implicitly defined within *CConvComparison* class declaration (i.e., they are inline functions), mainly because they are very short. Bigger member functions or constructors should be moved to implementation files, unless if they are heavily used (i.e., for optimization purpose). Therefore, it is recommended to later move the definition of constructor of *CConvComparison* class to the *ConvComp.cpp*. Notice *check4ValidConf()* was declared as having privacy specification of *private*, because it is exclusively called internally by other member functions of *CConvComparison*, i.e., it is not part of this class interface.

Similarly to C, a C++ program follows the same program skeleton on Listing 1.1 in Chapter 1, which consists of a collection of declarations, including classes, and member functions that begins executing with *main()*. Figure 5.6 presents two possible versions of a *main()*, with an object of *CConvComparison* created on the stack and on the heap by the uncommented and commented versions, respectively. Exactly as done before with built-in or native types, dot and "*point-to*" (a.k.a., member selection) operators are used to access and call the member function *run()* of *CConvComparison*, while for the pointer-based version, i.e., the commented *main()*, the *delete* operator is explicitly called to return back to the system the space reserved on the heap during the creation of the pointer-based object with the *new* operator.

Figure 5.6 Partial execution and state of main() till the creation of the object variable on the stack

Figure 5.7 shows that member functions are defined in the implementation file using the scope resolution operator (i.e., "*::*") like in *void CConvComparison::run()*, because C++ adds new scope rules specifically for classes. Thus, all symbols declared within a class have their own scope, which is different from those of external symbols, namespace names, function names, and other class names. Figure 5.7 also illustrates the new standard *include* format that comes without *.h*, e.g., *<iostream>* instead of *<iostream.h>*. The latter is still allowed only for backward compatibility with existing code, and it does not require the use of *using* directive, e.g., "*using namespace std;*".

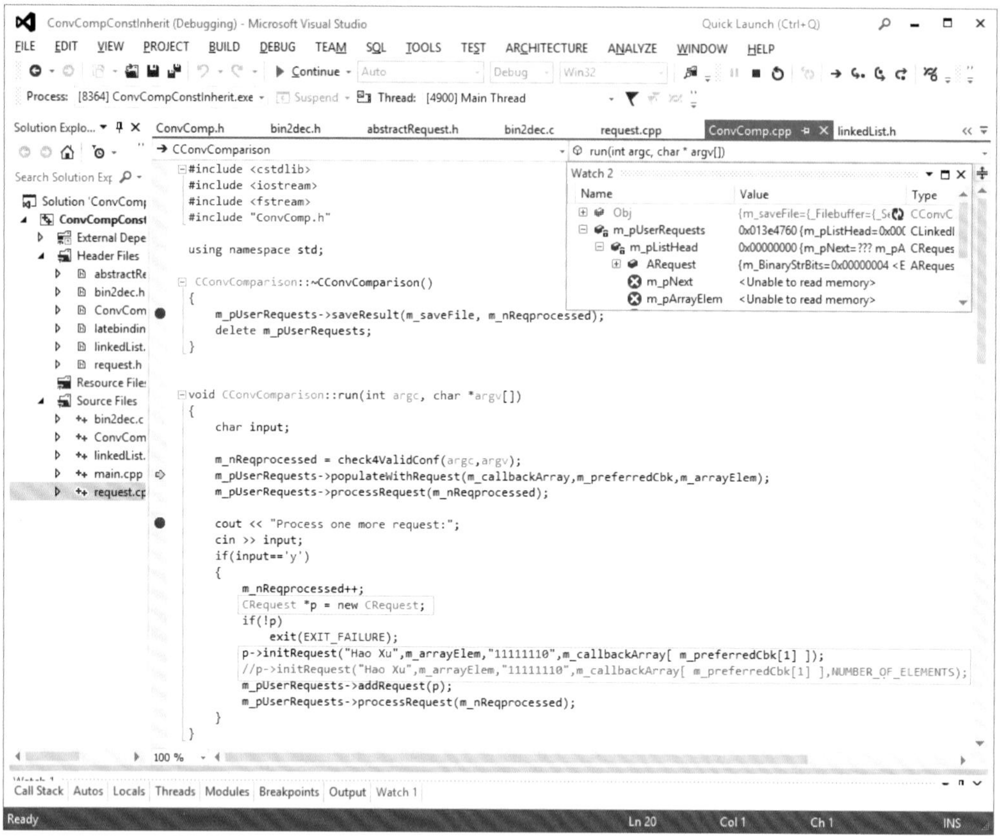

Figure 5.7   Partial execution of the member function CConvComparison::run() showing the initial setup of the linked list

The lowest blue-marked block of statements in Figure 5.7 presents two alternative calls to *initRequest()*, member function of *CRequest*, due to the default argument specified in its prototype (see the red-marked block in Figure 5.2). The specified default *NUMBER_OF_ELEMENTS* can be overridden by explicitly specifying another one, instead of itself as done in the commented statement. The upper blue-marked statement blocks of Figure 5.7 and Figure 5.8 exemplify on-the-fly variables definitions inside function and control construct bodies. Although C and C++ demand for "*definition before use of variables*", the former forces all variable definitions at the beginning of a scope.

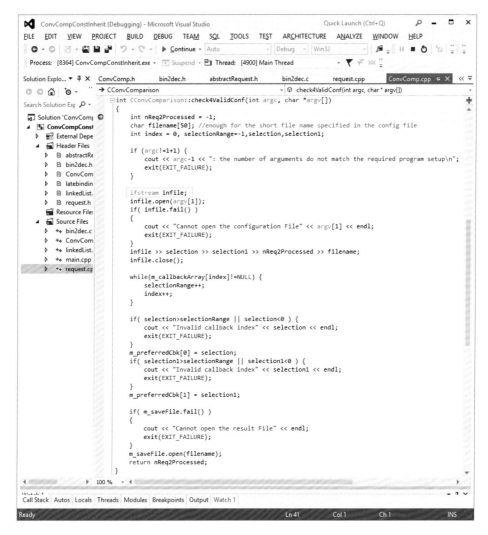

Figure 5.8   Starting the execution of the refactored member function CConvComparison::check4ValidConf()

The member function *CConvComparison::check4ValidConf()* is a refactored version of the C function presented in Figure 3.67, by replacing C standard file I/O functions with function members of classes *ifstream* and *ofstream*. Basically, objects of *ifstream* (e.g., *infile*) and *ofstream* (e.g., *m_saveFile*) are first created and then their equivalent to C standard C File I/O functions are called, using the dot operator.

Figure 5.9 to Figure 5.13 present the remaining implementations of classes which represent the binary comparison-conversion problem space. All member function are algorithmically equivalent to their C version discussed and coded in Chapter 4, with only minor differences explained below.

The blue-marked statement in Figure 5.10 illustrates how to call the base class member function, *ARequest::initRequest()*, from the derived class *CRequest* using the scope resolution operator. Notice that *CRequest::initRequest()* overrides *ARequest::initRequest()*, but with a different prototype.

Figure 5.9　Partial Implementation of CLinkedListRequest member functions

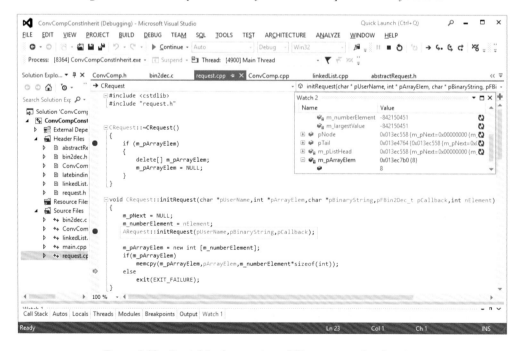

Figure 5.10　Partial Implementation of CRequest member functions

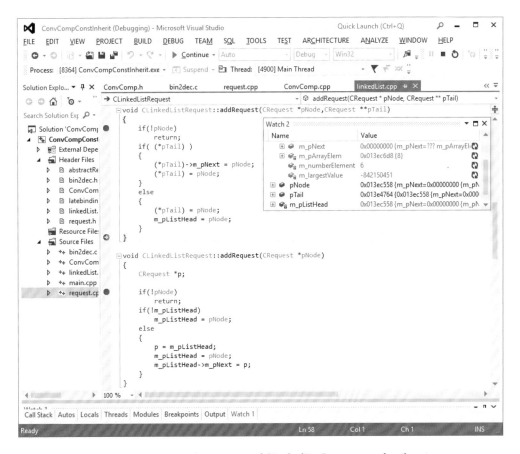

Figure 5.11   More Implementation of CLinkedListRequest member functions

Contrary to C where passing by reference is programmatically done using a pointer to the argument variable, C++ fully automate call by reference through the use of reference parameters, which are specified by prefixing the formal parameter name with an ampersand (&). The blue-marked statement in Figure 5.12 shows the prototype of the member function *CRequest::processRequest()* which specifies both formal parameters as references. The *CLinkedListRequest::saveResult()* in Figure 5.13 also specifies its first formal parameter as a reference parameter.

Looking back at Figure 5.2, the following question emerges: *why is the data member m_pNext part of the interface of CRequest*? The reason was making it easily available without performance cost to *ClinkedListRequest*, while promoting implementation reuse through composition than the reuse of the interface. Thus, another question will come to light: *why not provide corresponding accessor and mutator at the CRequest interface, as previously done with other heavily called data members from member functions of ClinkedListRequest*?

There is no apparent reason not to provide corresponding accessor and mutator *(a.k.a., set and get functions)*, but instead let's follow a different approach, by demonstrating the usage of class friendship, as shown by the uppermost blue-marked statements in Figure 5.14.

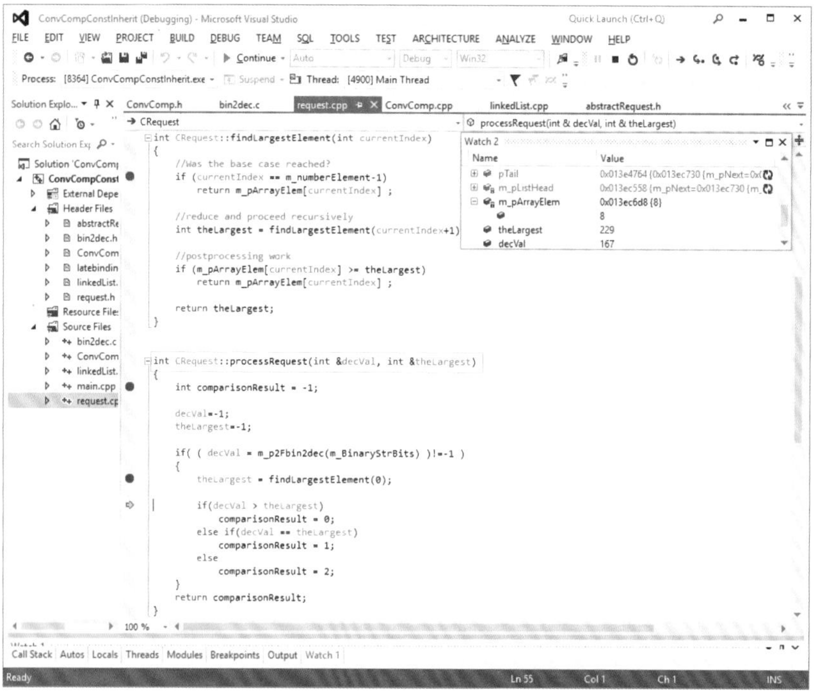

Figure 5.12   Remaining Implementation of CRequest member functions

Figure 5.13   Remaining Implementation of CLinkedListRequest member functions

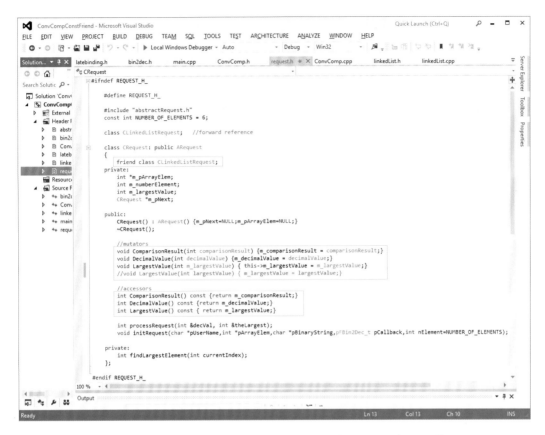

Figure 5.14 Promoting entire CLinkedListRequest class as friend of CRequest class to allows access to private member of CRequest

By declaring *CLinkedListRequest* as a friend of *CRequest through the* friend mechanism, the former is explicitly granted access to private members of the latter. The use of a forward declaration or reference, as shown in the uppermost blue-marked statement, is needed because *CLinkedListRequest* is referred inside *CRequest* before it is been declared. However, friend mechanism should be used with caution as it allows non-member function access to the hidden implementation of a class, and thus, escaping data-hiding restrictions of C++.

Strictly following OOP design, public members should in general be functions while data members should be private, in order to leverage the data-hiding principle. That is, data members are part of a class implementation and they need to be externally accessed through the class interface, by special member functions such as accessors and mutators. The two lowest blue-marked statements blocks in Figure 5.14 show the use of function overloading to provide the same name for both accessors and mutators, instead of declaring them by previously *get* and *set* functions. In doing so, reading-only or modifying the data members is determined by the way the overloaded function is called, i.e., with or without an argument. However, since accessors (i.e., the overloaded functions with no formal parameter) are read-only member functions, they must be declared *const (see the lowest blue-marked statements block).*

The overloaded mutator *LargestValue(int)* introduces you to the *this* pointer, which is an implicitly declared built-in self-referential pointer that can be used by any non-static member functions. That is, *this* keyword in C++ denotes an implicitly and automatically passed argument to all non-static member functions, each time they are called.

To eliminate the linkage specification to C code through *extern* "C", *findBin2Dec()* and *findBin2DecReverse()* are promoted as static member functions of *CConvComparison* class, as shown by the lowest blue-marked statements block in Figure 5.15. The implementation of the binary converters as previously presented in "*bin2dec*.c" will be moved to "*ConvComp.cpp*", with their function name prefixed with "*CConvComparison::*", like, for instance, in "*int CConvComparison::findBin2Dec(char * const binaryString) {····}*". Contrary to a non-static member, a static member of a class is independent of a particular object of its class, and it is externally accessed with the scope resolution operator. Furthermore, it cannot use the "*this*" pointer. The uppermost blue-marked statements block illustrates internal references to static member functions with and without the scope resolution.

Figure 5.15   Refactored version of CConvComparison to accommodate binary converters as static member functions

To conclude, let's shortly exercise the polymorphism mechanism, by refactoring *CLinked-Lis-tRequest::processRequest()* and *CLinkedListRequest::~CLinkedListRequest()* to manipulate *CRequest* objects through the on-the-fly defined base-class pointer, *pBase*. By accessing the virtual *ARequest::processRequest()* through *pBase* pointer variable, as illustrated by blue-marked statements in Figure 5.16, the appropriate function definition will be selected at runtime, in this case *CRequest::processRequest()*. Furthermore, the destructor of the class ARequest needs to be declared as virtual, i.e., as virtual *~ARequest()* in "*abstractRequest.h*", in order to call the overridden destructor of CRequest, and so, avoiding possible memory leaks. You can run the code in the debugging mode with inserted breakpoints at entries of *CLinkedListRequest* and *CRequest* destructors, for virtual and non-virtual destructor of *ARequest*, to see by yourself what will happen.

Figure 5.16 Refactored version of CLinkedListRequest member function and destructors to exercise the polymorphism mechanism

Previously in Figure 5.3 all symbols of *std* namespace were injected into the global namespace, as visible through the now commented red-marked statement in Figure 5.17. Using such a global *using* directive will be fine in an implementation file (i.e., a *.cpp* file), as the injection effect vanishes at the end of that file compilation. However, such practice is not recommended in header files because it easily pollutes the global namespace, leading to evil name clashes. Thus, the blue-marked statements in Figure 5.17 and Figure 5.18 show how explicitly and selectively qualified names can be used and imported instead, mainly because the only names that need to be visible, in this example, are the class *ofstream* and member functions *cin* and *cout*. Alternatively, occurrences of name clashes in implementation files can be easily managed by the programmer through restricted scoped using directives (e.g., inside individual function bodies such as in Figure 5.19), although it can harm the code readability.

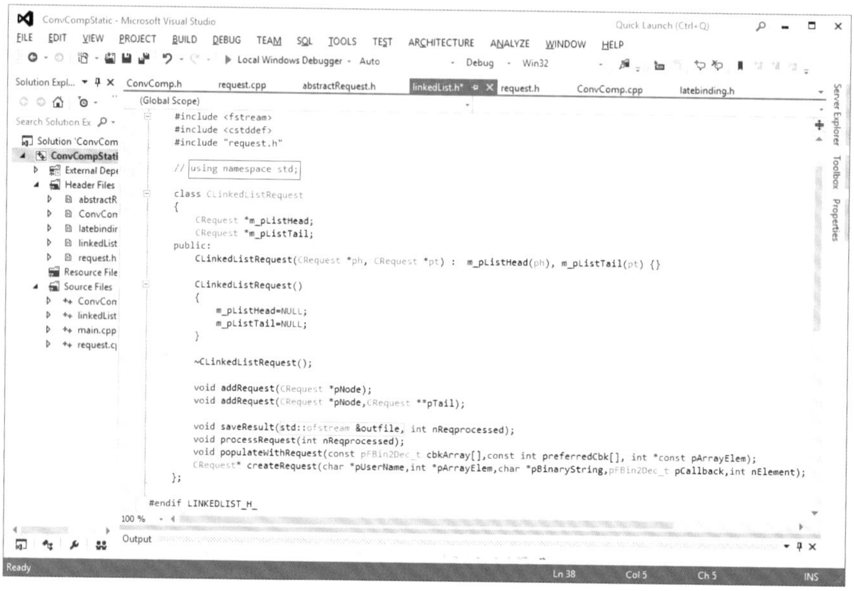

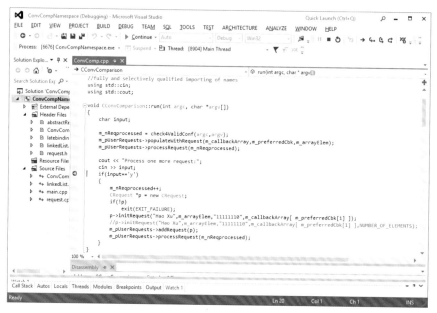

Figure 5.17   It is a bad programming practice a using directive in the global namespace as it can easily leads to evil name clashes

Figure 5.18   Namespace and fully qualified import of cin and cout names called by CConvComparison::run()

Figure 5.19   Namespace and restricted scoped using directive within CConvComparison::check4ValidConf() body

As pointed at the end of Chapter 4, there are some silent bugs in the above program that must be got rid of, as they lead to memory leaks. Those bugs will never manifest, unless a really huge amount of *CRequest* objects are made. A brief analysis of the program execution flow with inserted breakpoints at all "*exit(EXIT_FAILURE);*" statements on Figures 5.9, 5.10, 5.18 and 5.19, shows that the *Recommendation 30* described below, was not fulfilled. Let's start by presenting the new refactored version of *CConvComparison::CConvComparison()* shown in Figure 5.20, which control the dynamic memory allocation failure using new exception handling constructs of *try*, and *catch*. The C++ standard library provides *std::exception* defined in the *<exception>* header, as a base class specifically designed to declare objects to be thrown as exceptions. Among several exceptions thrown by parts of the C++ standard library, is *bad_alloc*, which is thrown by the *new* operator if an allocation failure occurs. Since C++ uses a termination model that forces the current blue-marked *try* block to terminate and transfers control to the red-marked *catch* block, in case of failure, the latter simply prints a proper message and then gracefully terminates the program. In our case, if you uncomment the allocation of the 5 blocks of integers, a *bad_alloc* exception will be thrown by the *try* block. Otherwise, you should add more blocks to force the occurrence of the exception.

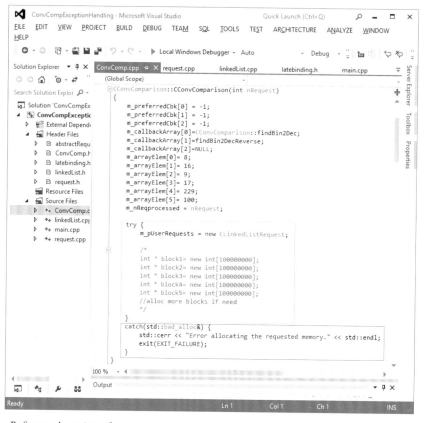

Figure 5.20   Refactored version of constructor CConvComparison() to control dynamic memory allocation failure using try block

Figure 5.21 shows the refactored version of *CConvComparison::run()* which monitors wrong command-line arguments and dynamic memory allocation failure, using *catch* blocks with different signatures, i.e., *catch(const char *msg)* and *catch(...)*, respectively. The blue-marked *try-catch* block is still monitoring and handling allocation failures that can occur while calling the following member functions: *CRequest::initRequest()* shown in Figure 5.22 and *CLinkedListRequest::createRequest()* illustrated by Figure 5.23. Notice they both rethrew the *bad_alloc* exception using the *throw* construct with and without arguments, which in these cases can be interchangeable, as they are simply identical. Both member functions rethrew the caught *bad_alloc* exception because their corresponding catch blocks cannot completely process them. In case of allocation failures during the execution of these member functions, rethrown exceptions will be caught by the lowest *try-catch* block of *CConvComparison::run()*. To complete their handling, the *catch-all* block performed some cleanup by deleting all previously allocated *CRequest* objects already stored on *CConvComparison::m_pUserRequests*, and thus, avoiding memory leaks. The catch all C++ exceptions was used only for demonstration purpose. Although it will catch any exception, no matter the type of the thrown exception, it is considered a bad programming practice and so, it is not recommended. Usually, it is replaced by a sequence of *catch* blocks, each one catching a specific exception. However, in this specific example, it could be simply replaced by only a "*catch (std::bad_alloc&)*". Since *CLinkedListRequest::populateWithRequest()* was called as a monitored statement from *CConvComparison::run()*, this latter member function will also catch all *bad_alloc* exceptions rethrew by *CLinkedListRequest::createRequest()*.

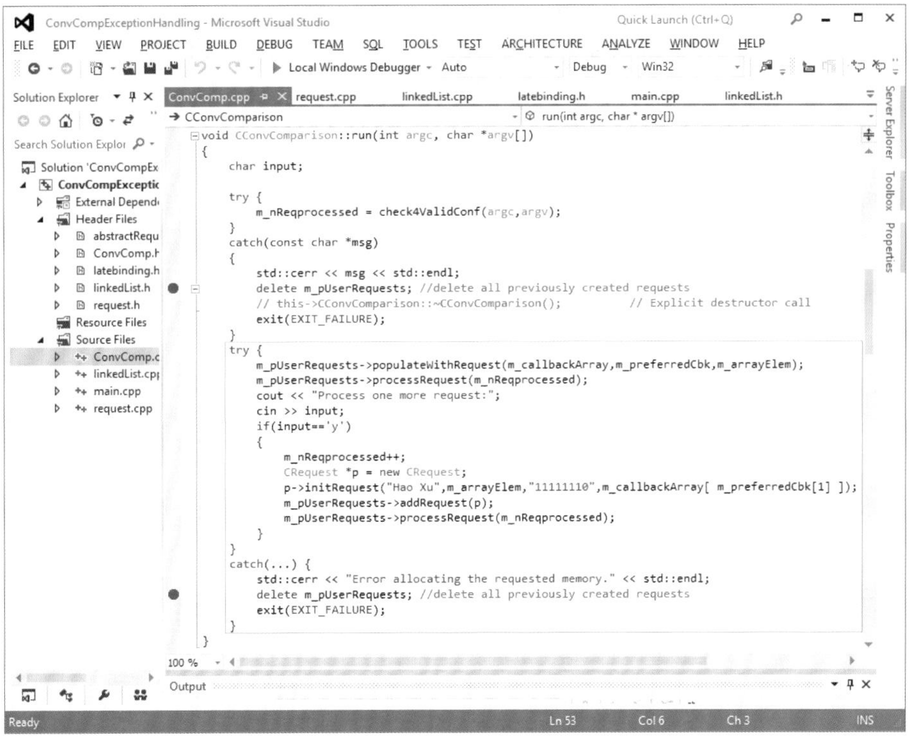

Figure 5.21   Refactored version of CConvComparison::run(), to monitor failures due to dynamic memory allocation failure and wrong command-line arguments

Figure 5.22   Refactored version of CRequest::initRequest(), to rethrew exception due to dynamic memory allocation failure

Figure 5.24 presents the new version of *CConvComparison::check4ValidConf()* refactored to control passed command-line arguments. The execution of this member function is monitored and handled by the uppermost try-catch block of *CConvComparison::run()* shown in Figure 5.21, with the catch block with a signature matching a constant string, i.e., "*catch(const char *msg)*". Some cleanup still needs to be performed before terminating the program, but so far no *CRequest* object was created. Only the empty *CConvComparison::m_pUserRequests* created on the heap by the *CConvComparison* class constructor needs to be deallocated.

Figure 5.23   Refactored version of CLinkedListRequest::createRequest(), to rethrew exception due to dynamic memory allocation failure

Figure 5.24   Refactored version of CConvComparison::check4ValidConf(), to rethrew exception due to wrong command-line arguments

All arguments of *throw* statements in Figure 5.24 are constant strings, matching the signature of uppermost *catch* block in Figure 5.21. To exercise the control of passed command-line arguments, firstly, insert breakpoints at all *catch* block as well as at the destructor of the class *CLinkedListRequest*. Secondly, run the program passing no command-line argument, a non-existing configuration file, and a configuration file with out of range callback indexes. Figure 5.25 illustrates execution results for the above scenarios.

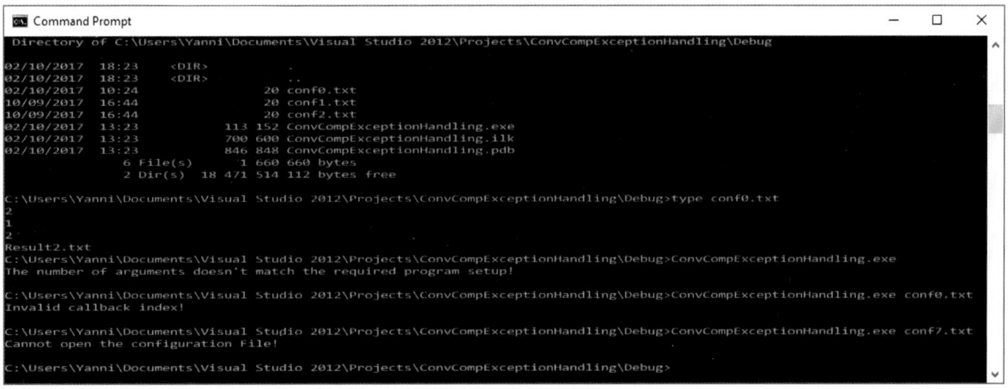

Figure 5.25   Execution results of the following 3 scenarios: (1) specifying no configuration file, (2) passing invalid callback index 2 at the first line of conf0.txt and (3) passing a non-existent configuration file

Finally, to evaluate yourself, comment the statement "*delete m_pUserRequests;*" and uncomment the statement "*this->CConvComparison::~CConvComparison();*" in the uppermost *catch* block of Figure 5.21, and then debug the program to pinpoint any problem, while fixing them all.

This chapter main focus was on porting C code to C++ code and it will end now with the following recommendations:

Recommendation 27: For readability, always name classes to clearly identify concrete from abstract ones. For example, we have been prefixed the name of a class with A and C for abstract and concrete classes, respectively. We also prefix data member names with "*m_*".

Recommendation 28: Always provide a virtual destructor to a base class with virtual methods to ensure proper cleanup of an object of its derived classes, when it is deleted through a pointer to the base class.

Recommendation 29: Linking code written in C through linkage specification mechanism is compiler-dependent and it can weaken the entire C++ type-safe model due to the C lack of type-safety. Thus, you should use alternative mechanism to unsafe linkage to non-C++ functions, such as static member functions, whenever possible.

Recommendation 30: To avoid memory leaks, always pair the usage of *new* and *delete* operators as done previously with *malloc()* and *free()* in C.

Recommendation 31: Do never use a global *using directive* into a header file to avoid losing the protection of that namespace, possibly polluting the global namespace and leading to evil name clashes. It exists to be used at the source file level where the programmer has full control of all included headers and their possible conflicts.

Recommendation SL: Accessor and mutators promote a safer design than directly accessible data members. Thus, always assign privacy specification of *private* or *protected* to data members, in order to leverage the key data-hiding principle of OOP and then provide accessors and mutators at class interface to externally access them. Since they are short functions that can be inline[①], no performance or code size impacts are noticeable.

References

[14] KELLY AL, POHL IRA. A book on C: programming in C. 4th ed. Boston: Addison-Wesley, 1998.

[15] SCHILDT HERBERT. *C++: The Complete Reference.* 4th ed. Berkeley California: McGraw-Hill, 2003.

[16] POHL IRA, C++ by Dissection. Boston: Addison-Wesley, 2002.

[17] GOYAL ARUNESH. Moving from C to C++. 2nd ed. New York: Apress, 2013.

Several other information sources were also used, mainly from internet, as well as those previously referenced in previous chapters. Thus, the credits also go to all of them.

---

① The code of inline functions are expanded in line at their call sites (i.e., at the point of their invocations).

Index

Access modifier 181
Accessor 23
ADT 182
Advanced 144
Algorithm 3
Alignment 107
Allocation 22
Analysis phase 3
argc 154
Argument 21
argv 154
Arithmetic 57
Array 3
Assembler 4
Base case 96
Base class 183
Breakpoint 25
Bug 4
Build process 4
Built-in types 31
Call-by-reference 70
Call-by-value 68
Callback 145
Callee 8
Caller 8
Call site 68
Catch-block 199
Class 182
Command-line 145
Compartmentalization 55
Compilation 4
Compile-time 19
Compiler 2
Composition 19
const 117
Constant 4

Construct 7
Constructor 183
CPU 7
C standard library 6
Customization 145
Data 2
Data-hiding 195
Debugger 5
Declaration 5
Declarative 6
Definition 5
Derived class 183
Design phase 46
Destructor 183
Divide and conquer 72
Directive 4
Do…While construct 43
Dynamic 202
Encapsulation 61
Environment 5
Evolution phase 56
Exception handling 182
Expression 6
File 4
Filesystem 145
Flowchart 52
For construct 41
free() 90
Friend mechanism 195
Function 3
Function pointer 145
Functional programming 6
Hands-on experience 181
Header guard 78
Header file 5
Heap manager 171

High-level	2
Global	7
Guard macro	79
IDE	5
If construct	36
Imperative	6
Implementation phase	46
Inheritance	182
Inheriting Class	185
Instance	3
Interface	62
Intermediate code	20
Interpreter	6
Iterative	2
Keyword	15
Late-binding	145
Lexical Analysis	20
libc	15
libm	15
Library	6
Lifecycle	144
Lifetime	8
Linkage	184
Linked list	170
Linker	2
Linking	5
Loop	7
Low-level	2
Macro	18
main()	8
malloc()	90
Member	32
Memory	21
Method	31
Middle-level	2
Modular	55
Modularity	61
Mutator	193
Namespace	89
Nibble	113
Object	4
Object-orientation	182
Object-Oriented Programming	182
Operator	23
Overloading	183
Parameter	7
Parameter passing mechanism	68
Parser	20
Pass-by-reference	70
Pass-by-value	68
Pointee	111
Pointer	2
Polymorphism	182
Pragma	79
Preprocessing	4
Preprocessor	4
Privacy specification	184
Private	182
Problem solution	2
Problem-solving	55
Problem statement	2
Procedural	6
Processor	4
Protected	145
Prototype	5
Pseudocode	55
Public	182
Qualifier	90
RAM	81
Recommendation	52
Recursion	96
Refactoring	135

Recursive	96
Recursive case	96
Reference	6
Return statement	46
Reusable	72
Reuse	2
Reuse of implementation	186
Runtime	6
Scale factor	112
Scanner	20
Scope	19
Segment	81
Semantic analysis	19
Semantic error	22
Setup	107
Signature	6
Source file	4
Statement	2
Static	6
Stream I/O	181
String	13
Structure/struct	31
Subroutine	58
Switch construct	37
Symbol	6
Syntax analysis	20
Syntax error	22
Testing and debugging phase	4
this pointer	196
Throw	184
Tuple	105
Try-block	184
typedef	32
Type-safe model	182
User-defined type	32
Variable	3
Variant	96
Virtual	82
Void	31
While construct	42
Wrapper	78